About Island Press

Since 1984, the nonprofit organization Island Press has been stimulating, shaping, and communicating ideas that are essential for solving environmental problems worldwide. With more than 800 titles in print and some 40 new releases each year, we are the nation's leading publisher on environmental issues. We identify innovative thinkers and emerging trends in the environmental field. We work with world-renowned experts and authors to develop cross-disciplinary solutions to environmental challenges.

Island Press designs and executes educational campaigns in conjunction with our authors to communicate their critical messages in print, in person, and online using the latest technologies, innovative programs, and the media. Our goal is to reach targeted audiences—scientists, policymakers, environmental advocates, urban planners, the media, and concerned citizens— with information that can be used to create the framework for long-term ecological health and human well-being.

Island Press gratefully acknowledges major support of our work by The Agua Fund, The Andrew W. Mellon Foundation, Betsy & Jesse Fink Foundation, The Bobolink Foundation, The Curtis and Edith Munson Foundation, Forrest C. and Frances H. Lattner Foundation, G.O. Forward Fund of the Saint Paul Foundation, Gordon and Betty Moore Foundation, The Kresge Foundation, The Margaret A. Cargill Foundation, New Mexico Water Initiative, a project of Hanuman Foundation, The Overbrook Foundation, The S.D. Bechtel, Jr. Foundation, The Summit Charitable Foundation, Inc., V. Kann Rasmussen Foundation, The Wallace Alexander Gerbode Foundation, and other generous supporters.

The opinions expressed in this book are those of the author(s) and do not necessarily reflect the views of our supporters.

The Carnivore Way

The Carnivore Way

Coexisting with and Conserving North America's Predators

by

Cristina Eisenberg

ISLANDPRESS

Washington | Covelo | London

Island Press is a trademark of The Center for Resource Economics.

The author gratefully acknowledges the contribution of maps by Curtis Edson and pen-and-ink drawings by Lael Gray.

Use of the quote on the epigraph page, from *Round River* by Aldo Leopold (1993), is by permission of Oxford University Press, USA.

Library of Congress Control Number: 2014933247

Printed on recycled, acid-free paper

Manufactured in the United States of America
10 9 8 7 6 5 4 3 2 1

Keywords: Island Press, carnivore, carnivore way, grizzly, grizzly bear, wolf, wolverine, lynx, cougar, jaguar, ecological effects, natural history, conservation history, rewilding, hunting, corridors, corridor ecology, spine of the continent, human fear, climate change, ecological benefits, trophic cascades, food webs, endangered species recovery

To my mentors, the late Tom Meier,
who encouraged me to study wolves and wildness,
and David Hibbs, who taught me about nature's tangled bank

To keep every cog and wheel is the first precaution of intelligent tinkering.

    — Aldo Leopold, *Round River*

Contents

Acknowledgments

Writing this book has been a journey, figuratively and literally, along the Carnivore Way. Stretching from northern Mexico to the Arctic Ocean, this landscape contains creatures with sharp teeth and claws living in the last redoubt of wildness in North America. As an ecologist, I'd long wondered what lay hidden along this path in the sort of places where a jaguar or wolverine might find refuge and prosper. Writing this book afforded me the opportunity to find out and to honor these creatures in stories and in science. I'm grateful to all the grizzly bears, wolves, wolverines, cougars, lynx, and jaguars who live along the Carnivore Way for your big lessons about wildness and why it matters in our rapidly changing world. Long may you run.

This work was inspired by the late Tom Meier, who directed the biological program in Denali National Park. He took me seriously when I came to him in the early 2000s—an inquisitive stay-at-home mom who had wolves running through her northwest Montana backyard and wanted to know what to do. He taught me to track wolves and encouraged me to go to graduate school and become a wildlife ecologist. Without his wisdom and generosity, this book would never have come to be.

I gratefully acknowledge the institutions that supported me in this endeavor, including Parks Canada, the College of Forestry at Oregon State University, and the Boone and Crockett Club. A generous grant from Wildlands Network, and Lissa and Paul Vahldiek, enabled much of the travel that made this book possible.

I had many mentors and guides as I wrote this book. It's impossible to properly thank all who helped, but I'll try. I'm grateful to my graduate advisor David Hibbs, who guided me in my doctoral studies and taught me to ask necessary questions and look at nature from multiple perspectives. I'm indebted intellectually to Walt Anderson, Paul Doescher, Tom Fleischner, Jerry Franklin, Norm Johnson, Bill Ripple, Doug Smith, and Michael Soulé. My thanks for their assistance with this book to Dan Ashe, Carita Bergman, Narcisse Blood, Bridget Borg, Janay Brun, John Burch, Doug Chadwick, Tony Clevenger, Mike Gibeau, Kerry Gunther, Jim Halfpenny, Patricia Hasbach, Steve Herrero, Jodi Hilty, Bob Inman, Peter Kahn, Kyran Kunkel, John Laundré, Kurt Menke, Oscar Moctezuma, Sue Morse, Alvine Mountainhorse, Michael Nelson, Lynne Nemeth, Carter Niemeyer, David Parsons, Laura Prugh, Charlie Russell, Hal Salwasser, Chris Servheen, Kelly Sivy, Kevin Van Tighem, John Vucetich, John Weaver, George Wuerthner, and Cynthia Lee Wolf. I'm grateful to John Russell of Waterton, Alberta, and Steve Clevidence and Dan Kerslake of the Bitterroot Valley, Montana, for teaching me about conserving carnivores in working landscapes.

A very special thank you to those whose compelling images grace and enrich this book. Curtis Edson created the powerful GIS maps that significantly inform the carnivore species chapters. Lael Gray created the evocative pen and ink drawings that accurately convey the nature of the large carnivore species profiled in this book. Thanks to Amar Athwal, Tony Clevenger, Steve Clevidence, John Marriott, Steve Michel, Brent Steiner, and Wildlands Network for providing photographs. I am grateful to Haida carver Jimmy Jones, for both the beautiful bear mask he created for me and the stories he shared. My thanks to Trevor Angel, formerly of Island Press, who during a Yellowstone pit stop on his epic round-the-world motorcycle journey took my book jacket photo.

Parks Canada conservation biologists and managers Rob Watt, Barb Johnston, Cyndi Smith, and Steve Michel helped me learn about relationships at an ecosystem scale, as did their American counterparts Jack Potter, Steve Gniadek, John Waller, Tara Carolin, Scott Emmerich, Regi Altop, Roy Renkin, and Rick McIntyre.

My thanks to the many conservation organizations and their dedicated staffs who helped with this book, including California Wolf Center, Center for Biological Diversity, Conservation Northwest, Defenders of Wildlife, EarthJustice, Grand Canyon Wildlands Trust, Living with Wolves, The Murie Science and Learning Center, Naturalia, The Nature Conservancy, National Wolf Watchers Coalition, UK Wolf Trust, WildEarth Guardians, Wildlands Network, Wildlife Conservation Society, Wolves of the Rockies, Yellowstone Association, and Yellowstone to Yukon. Jenn Broom, Mark Cooke, Jim Dutcher, Jamie Dutcher, Garrick Dutcher, John Davis, Elke Duerr, Wendy Francis, John Frederick, N. J. Gates, Karsten Heuer, Doug Honnold, John Horning, John Landis, Doug McLaughlin, Theresa Palmer, Denise Taylor, Chris Turloch, Kim Vacariu, Amaroq Weiss, and Louisa Wilcox were most helpful.

I thank the historians, librarians, and archivists who assisted my research in the following collections: Alaska and Polar Regions Collections and Archives at the University of Alaska, Denali National Park Archives, Glacier National Park Archives, The Murie Center, University of Wisconsin Archives, US Fish and Wildlife Service Archives at the National Conservation Training Center, and Waterton Lakes National Park Archives. Special thanks to historians and archivists Jane Bryant, Ann Fager, Mark Madison, and Rosemary Speranza.

A book like this wouldn't be possible without love and friendship. It had its genesis in the myriad questions that my daughters, Alana and Bianca Eisenberg, asked when wolves were recolonizing our land in the early 2000s. Heartfelt thanks to them and to Chris Anderson, Don Beans, Curtis Edson, my husband Steve Eisenberg, Donna Fleury, N. J. Gates, Valerie Goodness, Bill Jaynes, Laura Johnson, Kathleen Kelly, Sandy and Richard Kennedy, Kathy Ross, Leigh Schickendantz, and Brett Thuma for friendship, sustenance, and multiple reviews of this manuscript. My thanks to the late Patricia Monaghan, to Brian and Lyndsay Schott and all the Whitefish Review editors, and to Allison Hedge Coke, Andrew Schelling, Jonathan Skinner, Michael McDermott, and Black Earth Institute fellows for their advice on ecological literacy and tawny grammar. Big thanks to Dave Williams and Brenda Varnum for hosting me and my family in Banff, and to Lynne Nemeth for hosting me in Flagstaff and giving me a lovely place to write while there. Thanks to Tiffany Newman, Drake Doepke, and everyone at Saketome Sushi in Bigfork, Montana. Your friendship and delicious, healthy food nourished me at the end of many long writing days. My ecologist friends Trent Seager and Chris Langdon provided laughter and wisdom.

My field crews over the years have been awesome and have provided many lessons on trophic cascades and life. I am grateful to Chris Anderson, Nick Bromen, Steve Decina, Marcella Fremgen, Dan Hansche, Caleb Roberts, Sharon Smythe, and Edn Wynd. I wish you much success in your work as ecologists.

The wonderful people at Island Press brought this book to fruition. From start to finish, your vision and caring made working on this project a deeply fulfilling experience. My sincere gratitude to the incomparable Barbara Dean, who was instrumental in creating this book and helping me evolve as a writer. Managing the many details of publishing a book is an art. Huge thanks to Erin Johnson, who nurtured and shepherded this book expertly from start to finish, and to Sharis Simonian, Jamie Jennings, Julie Marshall, Jason Leppig, Meghan Bartels, and copyeditor Mike Fleming for helping me reach people effectively.

I thank my Carnivore Way traveling companions: my husband Steve, my niece and nephew Alisa and Andrew Acosta, and my friend Donna Fleury. None of them balked at the thought of putting in a 2,000-mile transect, and they followed me literally to the end of the earth. Their gifts of humor, vision, patience, and trust helped manifest this book.

Journey into Wildness

In December 1992, I picked up the latest issue of *National Geographic* magazine to do some light reading before bed. The cover article, by Michael Parfitt, profiled Highway 93—one of the toughest two-lane roads in North America.[1] I was a new mother back then, with seventeen- and six-month-old babies, born back-to-back. Between frequent feedings and diaper changes, most of the time I functioned in a sleep-deprived daze. My fuzzy mind had trouble absorbing much more than a magazine article now and then. I remember looking forward to sinking into that magazine.

Parfitt's article stopped me cold before I'd read a full page. He wrote about an anachronistic highway that ran through open spaces along the spine of the Rockies, from Arizona to Alberta. He described landscapes where both people and wildlife were unfettered. I wondered what it would be like to live in such places.

At the time, we lived on a ranch on the California Central Coast, where things were changing rapidly. In the previous five years we'd seen the human population there quadruple. Housing developments had sprouted on rolling hills that once held little more than oaks and tawny grasses. The last straw came when a Walmart and then a Costco popped up just down the road from our ranch. We knew we didn't want to raise our daughters here. We wanted to find a place that would be rural and a bit wild, with open space—somewhere that wouldn't change much over the coming years. Perhaps we were asking too much.

While I'd never seen the places Parfitt wrote about, his story clarified my thoughts about why our formerly rural California home no longer felt like a good fit. His article was about a road into wildness. He revealed enough about the mystery and wild wonders that awaited those who ventured there to make me want to go. I had a sense that we'd find what we needed on that highway. The article stuck with me, so much so that eighteen months later I found myself on Highway 93 in my Jeep, with our daughters and dog, on the way to find a new home. My husband, Steve, stayed behind to work.

And so I drove, taking my time and many side roads. In my quest I found golden canyons, trout-filled streams, and big, craggy mountains that plunged into a prairie sea. I also found clear-cuts, abandoned homesteads, and small towns that were expanding rapidly—just like home. The girls were good. As we traveled, the aspens and cottonwoods leafed out. We spent a beautiful Mother's Day in the Wyoming Tetons, picnicking next to an alpine lake where the ice had just broken up. The girls tried to find wildflowers to make me a bouquet, but they couldn't find any so early in the season. As I watched them play, I had a mother's insight about the rightness of our journey.

Steve had faith that I'd be safe and would find the right place. I was and I did, 7,500 miles later. When he flew out to northwest Montana to meet us, we took him to a narrow valley nestled between larch-clad mountain ranges. Creatures the likes of which hadn't been seen for decades in most of the United States still roamed this green, lush valley. The locals told us that the large carnivore population here outnumbered the human population. Steve agreed that it felt like home, and we've lived here ever since.

The land we bought had thick, second-growth forest and a large meadow on it. Our small, hand-hewn log cabin stood in that meadow. Undeveloped land surrounded ours, followed by a strip of state forest, a roadless hiking area, and the million-acre Bob Marshall Wilderness. Our closest neighbor lived a half mile away. The small, diverse community here included families with young children. The nearest town, which had a population of 1,000 people, lay ten miles away. In twenty years, none of this has changed, except now we have neighbors who live a quarter mile away and have become good friends, and the meadow on our land has become overgrown.

As we settled in, we learned about our land's natural history. We wanted to know everything about it—where the wild orchids bloomed in spring; where to find sweet, ripe huckleberries in late summer; how to coexist peacefully with grizzly bears; how to become better stewards of our forest. I became a writer and hunter.

In getting to know our land, I had powerful lessons. When I read *A Sand County Almanac* by Aldo Leopold for the first time, passages like "There are some who can live without wild things and some who cannot" inspired me to ask questions about what wildness means and why it matters.[2] In the late 1990s, when wolves recolonized our valley, I got some answers. White-tailed deer and elk that used to stand around complacently mowing aspens and shrubs down to ankle height had to stay wary and on the move to avoid being eaten by wolves. And within five years, unbrowsed aspens and shrubs had filled our meadow.

I became so curious about these relationships that when our daughters grew older, I went to graduate school and studied ecology. For my research topic, I chose large carnivores and food webs. I wanted to use science to show how the wildness that carnivores bring to a system touches everything and can create healthier ecosystems.

Recently, I reread Parfitt's article for the first time in twenty years. Three principal themes run through this article: wildness, corridors, and loss. A fourth theme bears mentioning: hope. He ends the article on a note of hope, with the image of a young woman purposefully striding away at the end of the road, her long, brown hair blowing in the wind. Since first reading this story, these four themes have permeated my life. I've sought out wildness, immersed myself in it, measured it empirically as a scientist, and helped others to understand and conserve it. For me, large carnivores embody wildness more than any other type of living creature. So, over the years, I've focused my work on large carnivore conservation.

The landscape Parfitt describes lies within the Carnivore Way. This corridor, which runs from Alaska to north-central Mexico, provides one of the last bastions of wildness in our fragmented world, as evidenced by the existence here of all the large carnivore species present at the time of the Lewis and Clark Expedition 200 years ago, which have been missing for the last 100 years from many places where they once roamed. From the beginning of my journey with my daughters, I learned to see this landscape the way Parfitt introduced me to it: on a continental scale, as a corridor where both humans and big, fierce creatures can make big movements.

The theme of loss has permeated my work, as has hope. As a scientist, I'm addressing ecological losses by helping to find ways to heal the damage we've done. Leopold wrote,

> One of the penalties of an ecological education is that one lives alone in a world of wounds. Much of the damage inflicted on land is quite invisible to laymen. An ecologist must either harden his shell and make believe that the consequences of science are none of his business, or he must be the doctor who sees the marks of death in a community that believes itself well and doesn't want to be told otherwise.[3]

An impressive body of science has shown us the importance of conserving large carnivores. For example, a recent article in the journal *Science* by William Ripple and his colleagues describes the massive, widespread global ecological impacts of carnivore extinction *and* restoration.[4] Indeed, our society has become so convinced of this that we have applied powerful environmental laws, such as the Endangered Species Act, and spent billions of dollars bringing wolves and grizzly bears from the brink of extinction. Leopold also wrote, "You cannot love game and hate predators." He called carnivore restoration "keeping every cog and wheel."[5] But we still have far to go. Now that we've recovered wolves and lifted their federal protection, we're in the process of reducing them to the lowest number possible above extinction. Grizzlies will be next. This is ecologically and morally wrong.

In writing *The Carnivore Way* I wanted to know: What are the biggest threats that large carnivores face as we move into the future? If humans are their biggest threat, how can we better coexist with them? And specifically, how can science, policy, and environmental ethics help us find a way to coexist? Answering these questions required telling the unique story of the large carnivore species along the Carnivore Way: their natural history, biological needs, vulnerabilities, and our human relationships with them. But it also required exploring this landscape in time and space following in the carnivores' footsteps.

Twelve thousand years ago, two ice sheets covered much of North America: the Laurentian Ice Sheet ranged over most of Canada, and the Cordilleran Ice Sheet blanketed the Pacific Coast. As these ice sheets receded, a path opened between them. This corridor gradually widened, creating a passage for plants and animals. In time, lush forests arose, and large mammals began to inhabit and move through the steep slopes and fertile valleys of the Rocky Mountains.[6] Today, animals continue to use this ancient migration and dispersal pathway. It's part of their ancestral memory; it's in their genes. Indeed, I like to think of it as a "carnivore way" because of the carnivores who've worn deep trails in this pathway since time immemorial.

The Carnivore Way begins in the Sierra Madre Occidental, 400 miles south of the US-Mexico border and ends in Alaska, at the shore of the Arctic Ocean. It contains the current range of the six large carnivore species in the West that are profiled in this book: wolves, grizzly bears, wolverines, lynx, jaguars, and cougars. I chose these species because they are the most imperiled of the large carnivores in this region.

Corridors function like lifelines, enabling animals to flow from one core area to another. Barriers to this basic need to move, such as human development, can provide formidable threats to long-term survival of many species.[7] For the large carnivores, it's not just about losing the freedom to move, it's about losing a natural process. They and other species use dispersal as a key survival mechanism to maintain genetic diversity. Species also use dispersal to adapt to climate change. Ten thousand years ago, when North America consisted of vast, unbroken tracts of land, these dispersals were probably fairly straightforward. But today, given our fragmented continent, such movements literally amount to acts of faith. Faith that by acting on instinct, they will find what they need to persist as individuals, and beyond that, as species: a safe home, suitable habitat, and a mate.

To learn more about the challenges carnivores face today, I traveled the Carnivore Way myself, building on the journey I took twenty years ago when I discovered our wild Montana home. My Carnivore Way travels included a 7,300-mile road trip from northwest Montana to Alaska. My husband Steve, my niece and nephew Alisa and Andrew Acosta, who were college students at the time, and my friend Donna Fleury, a science writer and naturalist, joined me on this journey. I also made trips to the Arctic, and from Montana to northern Mexico—another 6,000 miles.

In the far north, I discovered new depths of wildness and gained greater insights into our human relationships with the large carnivores. My early mentor Tom Meier, a wolf biologist who led the science program in Denali National Park, was among the many individuals who shared important lessons about coexistence. Grizzly, wolf, wolverine—the relationships people in the far north have with these species are more primal and spiritual, but simultaneously more sharply pragmatic, than similar relationships in the 48 contiguous United States. Perhaps this is because of the harshness and difficulty of survival here, or because these animals have never been fully eliminated, as in the south. This expansive northern landscape defies dewilding. Nevertheless, large carnivores face significant human threats to their well-being here, too.

My sojourn to the southwestern United States and northern Mexico taught me that carnivore conservation issues here are different from those of the far north. In the southern portion of the Carnivore Way, challenges include species reintroductions (e.g., wolf, lynx, and wolverine), the border fence, large cities, a lack of Mexican public lands, and a greater human footprint via cattle ranching and energy development. Drought and climate change further complicate survival for both carnivores and humans.

In this book, like the large carnivores that use the Carnivore Way, we'll cover much ground. In Part I: Wildways, we'll begin by exploring continental-scale corridors. I will share stories that illustrate the large carnivores' ecological function and why their existence is critical to our own. We'll look at what we need to be doing to assure carnivore health as well as the health of the landscapes they need in which to thrive. We'll consider carnivore conservation as a transboundary, international matter, because jaguars and grizzly bears don't stop at political borders. To that end, we'll compare environmental laws in Canada, the United States, and Mexico, as well as on Native American and First Nations sovereign lands.

In Part II: Where the Carnivores Roam, we'll explore the scientific, public policy, and conservation valence of each of the six carnivore species profiled in this book. We'll begin with their natural history and ecology. Because large carnivore conservation is ultimately about people, we'll review our historical and current relationships with these species. We also will look at tools to help them roam, which include problem-solving strategies such as collaboration.

So how *can* we create and support populations of large carnivores in our rapidly changing world? By learning to live more sustainably and ethically on the earth. Science and environmental law can help us learn to share landscapes with fierce creatures, but ultimately, coexistence has to do with our human hearts.

PART ONE

Wildways

CHAPTER ONE

Corridor Ecology and Large Carnivores

In early 2003, a two-year-old male lynx (*Lynx canadensis*) was cruising through his territory near Kamloops, British Columbia, searching for prey. As he maneuvered through the forest, padding easily in deep snow, he picked up the scent of food. He lowered his nose, took a step, and paused. All at once, what should have been just another in a lifetime of simple steps proved ill-fated. He found himself caught. No matter what he did, he couldn't free his paw from the hold of a trap. Soon a human came along and jabbed something sharp in his rump, which rendered him unable to move. When he came back to his senses, his life had changed in surprising ways.

Although the young lynx didn't know it, in addition to a bulky collar, he'd acquired a name: BC03M02. Wildlife managers and scientists had moved him to the United States for a lynx reintroduction program. Eventually he was released into southern Colorado's high country. It was much drier there than in the northern Rocky Mountain home where he was born, and there weren't as many snowshoe hares (*Lepus americanus*), his favorite prey. But he found a willing, fecund mate and enough to eat. Life was good; in two years he sired three litters of kittens. And then one day in late 2006, something in his brain, some inchoate longing, some homing instinct, made him feel like roaming. At first he simply traveled from one snowshoe hare stronghold to another, finding food when he needed it. After a while, he started ranging farther, and eventually just kept going. He ended up crossing landscapes unlike any he'd experienced before: the Wyoming Red Desert, followed by the Greater Yellowstone Ecosystem.

Colorado Parks and Wildlife (CPW) biologists determined that over the next several months, the lynx covered 2,000 miles. His last recorded collar signal before the battery gave out occurred in late April 2007. Eventually BC03M02 found his way remarkably close to where he was born, near Banff National Park, Alberta. And there his life ended, in another trapline—a lethal one set to legally harvest fur-bearing mammals. Superbly healthy at the time of his death, well-fed, with a luxuriant coat of fur, he set a world record for the greatest known distance traveled by one of his kind. Despite his tragic end, BC03M02 proved that even in our fractured world, it's possible for a carnivore to roam widely. But ultimately, the media hoopla about how far he'd traveled belied the tragedy of his death.

In recent years, lynx from the CPW reintroduction project also have dispersed south, into New Mexico's mountains. Researchers didn't anticipate these dispersals, which involved crossing interstate highways, traversing areas of high human use, and dodging death in myriad ways.[1] En route, these dispersers had to find snowshoe hares to eat—not always an easy task.

Other species besides lynx have an innate need to wander. These instinctive journeys involve both *migration*, defined as seasonal, cyclical movements from one region to another and back for breeding and feeding purposes, and also *dispersal*, the process animals use to leave their natal range and spread permanently from one place to another.[2] Many species, such as pronghorn (*Antilocapra americana*) and elk (*Cervus elaphus*), migrate as part of their annual life cycle. Fewer species disperse. Wolves (*Canis lupus*), wolverines (*Gulo gulo*), cougars (*Puma concolor*), jaguars (*Panthera onca*), lynx, and grizzy bears (*Ursus arctos*) have natural histories that often include long dispersals (although grizzly bears don't disperse as far as some of these other carnivore species).

In our developed world, landscape-scale dispersals are becoming increasingly challenging for all species, but they are still happening. In February 2008 in California's Tahoe National Forest, while being used to conduct a marten study (*Martes americana*), Oregon State University graduate student Katie Moriarty's remote-sensing camera photographed what appeared to be a wolverine. California's first substantiated wolverine sighting since the 1920s, the grainy image created a furor. Scientists began by trying to figure out where it had come from. The nearest known population was about 900 miles away, in Washington State. DNA tests of scat samples collected from the animal proved that it was genetically related to Rocky Mountain wolverines and had dispersed from an out-of-state population.[3]

In spring 2009, a two-year-old female Yellowstone wolf wearing an Argos satellite collar made an astonishing, 1,000-mile trek to north-central Colorado. Aspen-crowned mountains and deep valleys in that part of Colorado harbor abundant food for wolves: healthy deer (*Odocoileus hemionus*) and elk (*Cervus elaphus*) herds. However, it's not very safe to be a wolf in Colorado, due to low human tolerance for this species as well as other hazards. This wolf hung around Eagle County for a few weeks, but eventually died after eating poison intended for coyotes (*Canis latrans*).

One of the most compelling long-distance dispersals began in Oregon in September 2011, when a young male wolf left northeastern Oregon's Imnaha pack. Called OR7 by the Oregon Division of Fish and Wildlife (ODFW), he covered an astonishing range of territory through the state as he moved toward the southern Cascade Mountains. ODFW began posting images of his walkabout, which showed up on their maps as a thick, black zigzagging line. Popularly dubbed Journey, this wolf drew public attention as he wandered around Oregon in search of a mate. Often he looped and doubled back on himself, but each week he trended farther south. Just before Christmas 2011, he reached Crater Lake.

After Christmas, a public question arose: how far south would this vagabond wolf go? Some wondered whether he was looking for love in all the wrong places. Right before New Year's Eve, the public got their answer to the first question. Journey entered California, becoming the first wolf confirmed in the state since 1924. For the next fifteen months he stayed in Northern California and out of trouble. He preyed on deer and left cattle alone, but didn't find a mate. In mid-March 2013 he returned to southwest Oregon, where he remained through the end of that year. His California travels inspired the state legislature and Game Commission to consider putting the wolf on the state list of threatened and endangered plants and animals, and to begin creating a management plan for this species.

These transboundary stories make our hearts beat a little faster and give rise to complex emotions: wonder, grief, hope. How many other such dispersals of which we are unaware have there been? Why do so many have tragic endings? And how can we improve the outcomes, now that we're aware that these dispersals do occur?

Audacious as such dispersals may seem, some animals can't help making them. This behavior is imprinted in their DNA, in the shape of their bodies, and in how their minds work. Moreover, they do it with casual grace, as if such heroic dispersals amounted to just another day in their lives. There goes a wolf, loping a thousand miles in a harmonic, energy-conserving gait, its hind feet falling perfectly into the tracks left by its front feet. And a wolverine, effortlessly running up mountains and back down in minutes, covering more ground and elevation more rapidly than any other terrestrial mammal. If such behavior were a simple act of will, most of these stories would have happy endings. But nevertheless, these dispersals fill me with hope that perhaps we'll soon get it right, given the opportunities and powerful lessons these animals are providing.

The corridor that the carnivores use for dispersal in the West extends from Alaska to Mexico. It holds many stories and goes by various names. Long ago, the Blackfeet called it *miistakis*—the backbone of the world. The conservation organization Wildlands Network refers to this cordillera as the Spine of the Continent.[4] And I call this pathway worn into our continent by carnivore footfalls across the ages the Carnivore Way. Figure 1.1 depicts the Carnivore Way and the 2013 distribution of the six large carnivore species (grizzly bear, wolf, wolverine, lynx, cougar, jaguar) I write about that live within this corridor.

When I found our wild Montana home, I made a dispersal of my own that taught me to think about conservation on a landscape scale. My earliest lessons in our new home had to do with learning about the Crown of the Continent Ecosystem, arguably one of the most intact ecosystems south of Alaska.

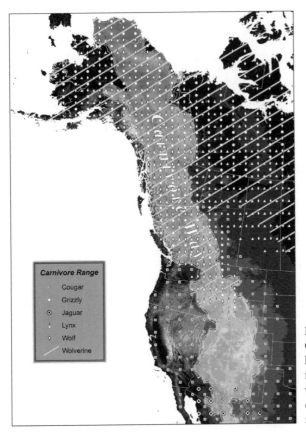

Figure 1.1. Map of the Carnivore Way, showing large carnivore distribution for six species: grizzly bear, wolf, wolverine, lynx, cougar, and jaguar. (GIS map by Curtis Edson.)

Getting to know this vast landscape, where mountains and public lands stretch from horizon to horizon, sparked my interest in *corridor ecology*—the science of how animals move across landscapes and inhabit them. I wanted to better understand the conservation requirements of animals that need to cover a lot of ground in order to thrive, and what people can do to create healthier, more connected landscapes for them.

The Backbone of the World

Our home lies roughly at the center of the Crown of the Continent Ecosystem, which contains some very big country. This 28-million-acre ecosystem extends east to where the Great Plains meet the Rocky Mountains, south to Montana's

Blackfoot River Valley, west to the Salish Mountains, and north to Alberta's Highwood Pass. The Continental Divide splits it north to south. Its jumbled terrain contains a crazy quilt of mountains: the Livingstone, Mission, and Whitefish ranges, among many others. Watersheds best characterize this ecosystem: the Elk, Flathead, Belly River, and Blackfoot. It contains a triple-divide peak from which rainfall flows into the Pacific, Arctic, and Atlantic oceans, and there are thousands of lakes of glacial origin. But anyone who has spent time here knows that the Crown of the Continent, its wildness seemingly endless, is far more than the sum of its parts.[5]

The large carnivores are part of what makes this place so ecologically remarkable. One of the two ecosystems in the 48 contiguous United States that contain all the wildlife species present at the time of the Lewis and Clark Expedition (1804–6), the Crown of the Continent provides critical habitat for animals that need room to roam. John Weaver, senior scientist for the Wildlife Conservation Society (WCS), has found seventeen carnivore species here, a number unmatched elsewhere in North America, including the Greater Yellowstone Ecosystem and Alaska.

According to Weaver, the Crown of the Continent matters for three principal reasons. First, because of its large, intact wildlands that provide more habitat security for carnivores than do other ecosystems with greater human development. Second, because of its connection to northern ecosystems with abundant large carnivores. Third, because of its physical and biological diversity. The Crown of the Continent has four climatic influences: Pacific Maritime in the west, prairie in the east, boreal in the north, and Great Basin in the southwest. Coupled with a tremendous range of elevation from prairie to peak, these influences create a variety of environmental conditions which, in turn, support a variety of ecological communities, from shortgrass prairie to old-growth rainforest.[6]

Within this ecosystem, rivers have incised narrow, fertile valleys into the mountains. These valleys—one of which I live in—represent the last bastions of wildness, where a grizzly bear can travel easily, finding abundant food and little trace of humanity. Yet even in a place as wild as this, much is at stake, because human encroachment places a great deal of stress on large carnivore habitat. For example, some of these animals are having their travel corridors cut off by logging, natural gas extraction, and backcountry recreation (e.g., the use of snowmobiles).

In the mid-2000s, I began doing research on wolves in Waterton-Glacier International Peace Park, which lies in the core of the Crown of the Continent. This peace park is composed of two national parks: Glacier National Park in the United States and Waterton Lakes National Park in Canada. Glacier, established in 1910 at the urging of New York naturalist George Bird Grinnell, comprises 1 million acres. Waterton, established in 1895 as a forest preserve, comprises 124,000 acres contiguous to Glacier. One hundred and seventy such peace parks exist worldwide, but Waterton-Glacier was the first. Dedicated to protecting biodiversity and natural and cultural resources, peace parks help maintain connectivity across boundaries. In 1995, the United Nations designated Waterton-Glacier a World Heritage Site.[7]

A closer look reveals why this is a critical linking landscape. At Logan Pass, the highest pass in Glacier accessible by car, the millions of visitors who stop here annually see protected lands stretching out in every direction. The hundreds of peaks all around make it obvious why the Blackfeet consider this the world's backbone.[8] Yet this area's wildness is not quite as big as it seems. Twenty miles away, the Blackfeet Reservation, which consists mainly of working ranches, forms Glacier's eastern boundary. The tribe has been fighting extensive natural gas exploration on their land, although various factions within the tribal community have differing opinions and interests.

Waterton-Glacier contains two principal wildlife corridors: the Flathead Valley and the Belly River Valley. Both cross international boundaries. The first begins in southeast British Columbia, where the Flathead River flows southward across the border into northwest Montana, forming the western boundary of Glacier National Park. This watershed provides an essential corridor for everything from grizzly bears and wolves to native westslope cutthroat trout (*Oncorhynchus clarkii lewisi*). Bull trout (*Salvelinus confluentus*) here make a dramatic 150-mile spawning journey from their Flathead Lake winter waters northward to their autumn spawning sites in the headwaters of the Canadian Flathead. But because this corridor contains rich deposits of coal, minerals, and coalbed gas, the potential for ecological damage by extracting these resources (e.g., mining and gas exploration that includes hydraulic fracturing, or "fracking") threatens the health of this corridor and its watershed. To keep it healthy and to protect its wild character from development, many people, including Canadian conservationist Harvey Locke and WCS senior scientist John Weaver, have been recommending expanding Waterton Lakes National Park into British Columbia.[9]

The second major corridor, the Belly River Valley, begins at the feet of Chief Mountain. This 9,000-foot massif, sacred to the Blackfeet, stands in Glacier, near the US-Canada border. From its headwaters, the Belly River flows northward into Canada, lending its name to the valley it carved out of limestone. One of the wildest places in the region, this valley was popular for trapping wolves, grizzly bears, wolverines, lynx, and cougars during the nineteenth-century fur-trading era. However, by the 1920s these and other fur-bearing mammals had been trapped out of existence. Later in the twentieth century, when these species gained protection in the United States and in the Canadian national parks, the Belly River Valley became one of the first places they returned. At the Chief Mountain port of entry, a granite obelisk and narrow clearcut demarcate the border crossing and provide a strong reminder that legal boundaries, however invisible to the large carnivores who cross them regularly, are very real in terms of land management.[10]

In Waterton Lakes National Park, the mountains abruptly meet the prairie, with no foothills to soften this knife-edged transition. A bunchgrass and wildflower sea slams into a wall of mountains. Waterton, small as it is (124,000 acres), represents an invaluable refuge for carnivores. The landscape that extends north along the Rocky Mountains past Waterton, toward the Kananaskis provincial wildland and Banff National Park, connects US populations of far-ranging species, such as grizzlies, to Canadian populations, helping to ensure their long-term survival. This connectivity makes the Crown of the Continent unique. In contrast, Yellowstone is an ecological island because human development has cut off many of its carnivores from other populations.

Immediately north of the US-Canada border, most large carnivore species are managed with scant (but growing) recognition of their contribution to international conservation. Beyond national parks, humans can kill most large carnivore species via hunting and trapping.[11] To use wolves as an example, in Alberta, a landowner is entitled to shoot a wolf at any time of year, without a license, within five miles of his or her private property or grazing-leased public land. The government permits baiting wolves with poison. Some Alberta townships still offer a wolf bounty. Despite all of the above, Alberta has a varying but well-established wolf population, with several packs ranging along the East Slope of the Rocky Mountains, from Banff to the US-Canada border.[12]

Waterton ecologist Barb Johnston has spent decades working with wildlife in this and other national parks, such as Banff, which is 1.6 million acres in size. When I asked her why a park as small as Waterton—which lies at the other end of the size spectrum from Banff—is so important to landscape-scale conservation, she said, "Usually when you talk about carnivores, the main theme is large, undeveloped space that they can roam in. Waterton doesn't fall into that category, because it is so tiny. But because it is positioned where it is, we are part of a travel corridor for carnivores. Waterton is adjacent to source populations for carnivores, particularly grizzly bears and wolves, but also wolverines. We are a critical linking land because we are also adjacent to what makes population sinks for species that don't do so well in human-dominated areas."[13]

Outside Waterton, wilderness meets the Anthropocene. Coined by ecologist Eugene Stoermer in the 1980s, but popularized by Dutch chemist Paul Crutzen in 2002, the term *Anthropocene* refers to the geological epoch that began with the Industrial Revolution in the late seventeenth century. Ranchlands, developed over a century ago for cattle production, bound Waterton to the east. And 200 miles north lies Calgary, one of Canada's largest cities. Banff National Park, the most visited Canadian park, lies 65 miles west of Calgary, along Highway 1. This four-lane road, also known as the Trans-Canada Highway, slices through the heart of Banff via the Bow Valley. Completed in 1965 to connect the Canadian interior to the West Coast, this originally low-volume, two-lane highway once had little impact on wildlife. In 1950, Calgary had 130,000 people; by 2011, it had over 1 million. Also by 2011, approximately 30,000 vehicles passed through the Bow Valley daily. Such human impacts, which bring with them an accelerated extinction rate, typify the Anthropocene—the Age of Man.[14]

Epic Treks, Epic Visions

If any individual animal can be seen as the standard bearer for early corridor ecology, that animal would be the young female wolf from Paul Paquet's Kananaskis study area. He named her Pluie (French for "rain"), after the downpour in which she was collared in 1991. For the next eighteen months, a young warden named Karsten Heuer tracked her with an antenna and receiver as she blazed a remarkable trail. She ended up traveling from Alberta across the US-Canada border into Montana and Idaho, back to Alberta, and into British Columbia. By the time a hunter legally shot her in 1995, she'd crossed more than 30 legal jurisdictions and had rambled over an area of more than 40,000 square miles, double the size of Switzerland.[15] Banff National Park was but part of Pluie's range. In her travels, she demonstrated that parks, even very large ones, aren't enough to conserve species that need to travel widely.

Pluie's journey inspired Canadian conservationist Harvey Locke to think about landscape conservation at much larger scales. Locke's ideas caught on like wildfire and got scientists like Michael Soulé and Reed Noss thinking about continental-scale conservation. In 1993, conservation leaders and scientists met at a Kananaskis research station and created the Yellowstone to Yukon Conservation Initiative (Y2Y). Conceived by Harvey Locke, president of the Canadian Parks and Wilderness Society (CPAWS) and board member of the Wildlands Project (now known as Wildlands Network), Y2Y was a joint CPAWS and Wildlands Project effort. Wendy Francis, a Y2Y leader present at this founding meeting, says, "The 'aha' moment was the awareness that conservation can't be just about creating more protected areas. It has to be about bears and other animals that are going to have to live on the land outside the parks." Inspired by Pluie's movements, they envisioned a conservation corridor from the Wind River Range in western Wyoming to the Peel River north of the Mackenzie Mountains in the Yukon and Northwest Territories.

Today Y2Y focuses on large carnivores and on maintaining this corridor's ecological processes. A joint Canada-US effort that includes multiple non-profit organizations, Y2Y works with local communities and takes a scientific approach to conservation. Twenty years since its inception, Francis says, "One of the tremendous benefits and outcomes of Y2Y is that we have this network across the international boundary that never existed before."[16]

I visited Banff to learn more about corridor ecology. While there, I spent time with Heuer, a friendly, modest man with curly blonde hair and a boyish grin. President of Y2Y since 2013, he started to put down roots in the region when he radio-tracked Pluie in the early 1990s. In 1998, he made an epic, 2,200-mile, eighteen-month trek from Yellowstone to Watson Lake, Yukon Territory. When I asked what inspired this trek, he explained,

> When this concept of Y2Y started coming up, it was really powerful. . . . I just wanted to do something. And so I decided to find out if this thing that I was so excited about was actually workable on the ground. That was the birth of the idea to trek—to try to be one of those animals, a grizzly bear or a wolf or wolverine, trying to make these movements from one reserve to another, and see not just the physical barriers that might exist, but also the political, economic, and value-based ones among the people.[17]

To illustrate what he was finding, Heuer needed an indicator of connectivity. An *indicator species* is one whose presence provides a gauge of the overall health of an ecosystem. By monitoring indicator species (i.e., measuring a species' population and its ability to meet its basic biological needs: to find food and reproduce), scientists can determine how changes in the environment are likely to affect other species. Heuer used grizzly bear presence as an indicator of connectivity. While "trying to be a grizzly bear" during the 188 days of his trek, he found this species' scats, tracks, and rubs 85 percent of the time. Perhaps his strongest take-home message was that grizzly presence along so much of his route demonstrated that the Y2Y vision wasn't about creating something new. It was about acknowledging and honoring something we already have—landscapes that are connected and are providing passageways for animals that need to cover a lot of ground. This acknowledgment included a greater awareness of the fact that roads are one of the biggest barriers to connectivity for wildlife. In his trek, Heuer helped make us aware of places where an animal like the grizzly bear might have trouble crossing a highway to find a mate, for example, and where it would not take much to mend those gaps.

Following in Heuer's footsteps (sometimes literally), writer, naturalist, and Wildlands Network cofounder John Davis made some continental treks of his own. In 2011, he completed TrekEast for Wildlands Network, a 7,600-mile muscle-powered exploration of the wild corridor from Florida to southeastern Canada. Then in January 2013, Davis set out from Sonora, Mexico, to traverse the Rockies as far north as southern British Columbia. He ended up walking, biking, and paddling 5,000 miles in eight months. Along the northern portion of this trek, he followed Heuer's path—and Heuer himself joined him for the last leg of the journey.

Wildlands Network's objectives for TrekWest were to ground-truth the Spine of the Continent. Along the way, Davis and conservation partners illuminated the issues faced by sensitive species—especially large carnivores—that need room to roam long distances. In doing so, they strengthened relationships between outdoor recreationists and conservationists, and did outreach in communicating wildlife needs. When I asked about his most powerful lessons, Davis said,

> I did these treks in part to get a visceral feel for what it means to connect wildlands and to see if it's still possible to protect nature on a continental scale. What I found was that it may not be as easy in the West as in the East, due to ranching. We have a lot of growing to do in terms of partnerships with people in the West. We also need safe crossings for wildlife, and perhaps that's where we can find the most common ground. But, by far, the toughest issue is livestock grazing on public lands. This complex issue divides the conservation community. I think the best way to find common ground is to get people outdoors, to connect them physically to landscapes.[18]

Continental-Scale Wildlife Corridors

Conservation biology is the science of protecting biodiversity. Soulé and others concerned about widespread habitat degradation and species extinctions founded this new science in the mid-1980s. A lean, ascetic-looking man in his seventies, Soulé has been my friend and mentor for several years.

One of my favorite Soulé moments occurred some years ago when he joined me and our field crew on our aspen (*Populus tremuloides*) research project on the High Lonesome Ranch, a 400-square-mile conservation property in Colorado. Deep in an aspen stand where we were examining forest conditions, we broke for lunch. As he sat cross-legged on the ground, the crew, which mainly consisted of students in their early twenties, surrounded him, eagerly awaiting words of wisdom. Instead, he started quizzing them about ecology, asking questions with no easy answers, sounding more like a Zen master than a scientist. When a crew member asked him to define the science of conservation biology, he simply said that it's the science of saving nature.

In the early 1990s, Soulé cofounded The Wildlands Project, based on the idea that to preserve the diversity of life on Earth, we need to create permeable corridors where species can travel. But not just any corridor would do. According to Soulé,

> Many large animals move hundreds or thousands of miles in their lifetime. Those journeys are not only poetically or epically significant, but biologically significant, because it means that those creatures use and need large areas to survive and reproduce. The Wildlands Project created a consciousness of scale. Not only a local scale, where we live, where creatures have their home ranges, but a regional scale and a continental scale over which creatures may move during their lifetimes.[19]

More than twenty years later, this vision has matured, in part due to work by Wildlands Network conservation partners Y2Y and WCS, to address habitat fragmentation created by human development and climate change.[20]

Jodi Hilty, who leads the WCS North American Program, helped pioneer corridor ecology. I met with her between snowstorms on a wintry afternoon in Bozeman, Montana. When I asked her to identify the most critical corridors for large carnivores in the Rocky Mountains, she said, "The Spine of the Continent is *the* important corridor. So I think the question is actually: What are the bottleneck places that we need to ensure stay connected?"[21]

Hilty explained that the far north was still relatively intact. However, much was at stake there due to increased development in the Yukon. She expressed hope that conservation planning will lead to carefully thought-out protected areas in northern watersheds such as the Peel River in the Yukon and northern British Columbia, which will help sustain the largest wild carnivore populations in North America. In the Canadian Rockies she identified two major highway bottlenecks that disrupt north–south wildlife movements: the Trans-Canada Highway Corridor in Banff, one of the busiest thoroughfares in North America, and Crowsnest Pass on Highway 3, just north of Waterton. For years, expansion of Highway 3 into a divided highway (called *twinning*) has been under discussion. In its current two-lane state, Michael Proctor and his colleagues have detected no female and only two male grizzly bears crossing this road. These findings graphically illustrate the lack of connectivity for large carnivores on this highway.[22] Lack of adequate protected areas (i.e., places that governments keep as roadless and undeveloped as possible, and where humans can't hunt animals) in the enormous area between Banff and Waterton Lakes National Parks is also a long-term challenge, because this has created divided populations of animals that now have trouble fulfilling their needs to find food and a mate.

Various people and organizations are working to address these problems. Their independent studies complement one another and are adding to our growing picture of why maintaining connected landscapes matters if we want healthy populations of animals such as grizzly bears, and where problem areas lie, and what we can do to fix these problems. Many of the areas identified by Hilty are included in this international effort. While not a formal collaboration, these studies fit together to help us figure out where and how to improve connections for wildlife.

John Weaver has studied the Crown of the Continent Ecosystem, from Highwood Pass to the Blackfoot River, since 1985. His research has focused on corridors and population cores, and on the conservation status, habitat needs, and resilience of the grizzly bear, wolverine, and other sensitive species that need connected landscapes. These other species include bull trout, westslope cutthroat trout, mountain goat (*Oreamnus americanus*), and Rocky Mountain bighorn sheep (*Ovis canadensis*). Weaver has found protected lands, particularly roadless areas, especially vital to these species' well-being. Trout are vulnerable to logging, mining, and energy extraction, which increase silt, contamination with heavy metals, and water temperature, making streams uninhabitable. Additionally, the introduction of nonnative trout has contributed to the demise of native trout. Mammals are vulnerable due to the roads that development brings: roadkills reduce wildlife population growth and gene flow between subpopulations. Consequently, many species, such as the grizzly bear, avoid roads.[23]

Other scientists studying corridor ecology in the Crown of the Continent include Clayton Apps, Bryce Bateman, Bruce McLellan, Paul Paquet, and Michael Proctor. They recommend creating wildlife crossing structures along Highway 3, approximately 35 miles north of Waterton, in order to keep US populations of carnivores connected to Canadian populations. In 2010, the Western Wildlife Transportation Institute, Y2Y, and the Miistakis Institute prepared a Highway 3 report to help *mitigate* (i.e., alleviate) connectivity problems for wildlife. This plan is being used on the Alberta side of Highway 3 to install wildlife underpasses at Rock Creek and fencing to keep sheep off the road at Crowsnest Lakes.[24]

More recently, in his WCS *Safe Havens, Safe Passages* report, Weaver noted that habitat security and quality for wildlife has declined in this region. For many years, low road density and modest levels of natural-resources extraction gave this region *de facto* protection. However, road density has now grown to accommodate coal and timber extraction. Further, climate change has brought longer, drier summers, reducing snowpack and stream flows and increasing severe fires. This means animals must roam more widely to meet their needs.[25]

Weaver analyzed this area's ability to meet the habitat and conservation needs of what he calls the sentinels of the southern Canadian Rockies (the grizzly bear, wolverine, mountain goat, bull trout, and westslope cutthroat trout). He found sixteen mountain passes that provide critical connectivity across the Continental Divide between Alberta and British Columbia. Highway 3 represents a particularly problematic corridor due to its high volume of vehicle traffic. In sum, Weaver recommends designating 1.8 million acres in British Columbia as the Southern Canadian Rockies Wildlife Management Area. This area would be managed to conserve wildlife and minimize/mitigate human impacts on its habitat, by, for example, installing wildlife-crossing structures such as tunnels and overpasses and by limiting construction of new roads.[26]

In the northern Rockies, corridor ecologists including Proctor and Hilty have found additional landscapes needing protected connecting corridors for carnivores. These corridors would be created via wildlife-crossing structures and by moderating traffic, new development, and natural-resources extraction (e.g., natural-gas development). To the west of the Crown of the Continent, these include the Cabinet, Purcell, and Selkirk Mountains.

Hilty has identified other corridors at risk in North America. The High Divide, the land surrounded by the Crown of the Continent Ecosystem, the Salmon Selway region, and the Greater Yellowstone Ecosystem, links these three core habitats and is an important corridor for large carnivores like grizzly bears. In Wyoming, natural-gas extraction threatens wildlife movements from mountainous summer habitats to sage-grassland winter areas, particularly for pronghorn. Specific threats include new roads put in to support natural-gas extraction and the accompanying high volume of semitruck traffic and increasing human infrastructure (e.g., housing for workers). Farther south, according to Hilty, "From a wolverine's perspective, Colorado has a big chunk of unoccupied habitat. The same goes for a number of other species. How much connectivity there currently is between the Greater Yellowstone Ecosystem and Colorado and how to secure it are big questions." She also notes that the US-Mexico border, particularly since installation of the massive border fence, creates connectivity issues that affect many species, including the jaguar.[27] In the chapters that follow, we'll be looking, species by species, at these landscapes and their connectivity issues.

The Banff Wildlife Crossings Project

Banff, the crown jewel of the Canadian national parks, is a paradox. Big, primeval wildness and all it embodies—peaks, waterfalls, and glaciers—surround the town site, which is the antithesis of wildness with its bustling boutiques and hotels. All this activity brings people. To reduce roadkill in the busy Bow Valley and to create a more-connected landscape for wildlife, the federal government and Parks Canada fenced the 50-mile corridor between Canmore and Lake Louise along Highway 1, and built many wildlife-crossing structures. The fencing prevents wildlife mortality on the highway, and the crossing structures allow the connectivity that the fencing precludes. Trail cameras and radio collars have documented over 160,000 individual wildlife crossings in the last seventeen years, so the structures are working well. Other notable highway-corridor ecology projects, such as the Highway 93 mitigation in Montana and the Tijeras Canyon Highway 40 mitigation in New Mexico, offer safe crossings for wildlife. I looked at the Banff Wildlife Crossings Project as an example of such a study. In place longer than the other projects, it's enabling us to begin to see the long-term benefits of crossing structures for wildlife.

Since the 1990s, corridor ecologist Tony Clevenger has been guiding the Banff Wildlife Crossings Project. Animals used to die frequently when crossing the highway in the Bow Valley. According to Clevenger,

> In the 1980s, elk roadkill was outrageously high—Parks Canada averaged about 100 elk killed per year on that stretch of highway. They called it the "meat maker." And originally, it was. The government wanted to twin the highway and basically just fence it to keep elk out. But there was pressure to put in some underpasses, because fencing the highway was going to isolate elk on both sides.

The initial focus was mainly elk, because back in the 1970s, carnivores weren't protected in Canadian national parks, and outside national parks elk were an important economic resource via hunting.[28]

The Canadian government proposed an ambitious staged highway twinning and mitigation project that was eventually implemented over a 30-year period. During the decades of the Banff Wildlife Crossings Project, due to developing technology, corridor ecology science, and park policy regarding carnivores, the federal government planned some aspects of its phases incrementally. As we shall see, some phases represented shifts in awareness of which animals needed to be protected and how they might find safer ways to cross the highway.

The Banff Wildlife Crossings Project has consisted of the following: Phase I, completed in 1985, covered seven miles from the Park's East Gate to the Banff town site. Wildlife-vehicle collision-mitigation measures included an eight-foot-tall wildlife exclusion fence on both sides of the highway and six wildlife underpasses. Phase II, completed in 1988, extended the mitigation seventeen miles from the Park's East Gate, creating another four wildlife underpasses. Phase IIIA, completed in 1997, added 28 miles and 12 more wildlife-crossing structures, including the first two 50-meter-wide overpasses. Other improvements included a buried fence, which reduced the likelihood that species such as wolves will tunnel under the fence. (Wolves—notorious diggers—would often tunnel under the fence and end up on the highway, where they were sometimes hit by vehicles. Other species don't have quite the same compulsion to dig as do wolves.) Phase IIIB, recently completed from Castle Junction to the British Columbia border, brought the total number of mitigation structures to 60 (39 underpasses, 6 overpasses, and 15 culverts).

Of these phases, Phase IIIA best illustrates how managers began to apply emerging findings about carnivore travel-corridor needs. Initially, Phase IIIA was envisioned merely as more underpasses and fencing. But by the time this phase was being planned, Parks Canada had undertaken a mission to conserve carnivores. And so, because of the need to maintain connectivity across the highway for large carnivore species, scientists such as Banff grizzly bear ecologist Mike Gibeau, as well as conservation groups such as CPAWS, the Bow Valley Naturalists, and Y2Y, implemented a public campaign to include overpasses in Parks Canada's design. They'd learned about some of the overpasses in Europe, which larger-bodied species such as grizzly bears would be more likely to use than underpasses, so they pushed for two such structures—and got them. The park simply positioned the Wolverine and Red Earth Overpasses, as they were called, in areas with the highest number of ungulate (hoofed animal) roadkills. At the time, the relatively new Banff wolf and grizzly bear science programs had insufficient data on these species' movements to identify optimal crossing locations for them. To reduce construction costs, other Phase IIIA wildlife crossing structures were placed mainly near old borrow pits (excavated areas where material has been dug to use as landfill) that had reverted to meadows.[29]

C38O

In Banff, I visited some of the crossing structures with Steve Michel, a Parks Canada human-wildlife conflicts officer. We pulled over on Highway 1, west of the Banff town site, at the paired Wolverine Overpass and Underpass. Planners hypothesized that different types of wildlife might prefer different types of crossing structures (i.e., some preferring the shelter of an underpass versus the openness of an overpass). Therefore, by pairing an overpass with an underpass, it might be possible to enable more types of animals to cross.

Through a gate in the fence, Michel took me into a haven-like world. Outside the fence, vehicles hurtled by—everything from motorcycles to semitrucks that shook the ground with their passing. But as we walked onto a grassy apron to the top of the overpass, the deafening traffic noise subsided. We'd entered a world far safer for wildlife than the pavement and metal below. This inviting living landscape had multiple levels of native vegetation, game trails, and abundant wildlife sign—piles of elk pellets and berry-laden bear scats. Michel pointed out a trail camera aimed at one of the more prominent wildlife trails, positioned to collect data about what animals were using the overpass. He deftly evaded the camera as he guided me back down (fig. 1.2).

Figure 1.2. One of the Bow Valley Wildlife overpasses across Highway 1, part of the Banff Wildlife Crossings Project in Banff National Park, Alberta. (Photo by Steve Michel, Banff National Park.)

Next, we clambered over shrubs and deadfall within the fence and explored the paired crossing structure located 200 yards west. An elliptical concrete tunnel 13 feet high and 23 feet wide, the Wolverine Underpass was one of the smaller wildlife crossing structures on this highway. Michel explained that box culverts were even smaller. As I entered the underpass, the traffic noise receded, and I gasped in wonder. What looked like hundreds of deer and elk tracks, pressed into the bare earth inside, provided abundant testimony to this structure's effectiveness. The dim, slanting light that entered the underpass gave the tracks an ancient, fossilized appearance.

Initially, people didn't think wildlife were using the structures. Some objected to millions of dollars of taxpayer funds being spent in this manner. Clevenger considers monitoring and assessment essential to any highway wildlife-crossing project and has turned keeping track of what animals use these structures into a science. When the park first hired him in 1996, staff hadn't been documenting wildlife use methodically. Clevenger implemented a system of plots that contained soft dirt, called *track pads,* that he and his team checked regularly. More recently, he added motion-sensitive cameras. Through such means, he found differential use of the structures. For example, grizzlies preferred overpasses; cougars preferred underpasses; and wolves and moose preferred larger, more open structures (figs. 1.3 and 1.4). He also found that human activity at or near the crossings posed the strongest deterrent to wildlife use.[30] Clevenger explains,

> After the first five years of monitoring, we discovered there were thousands of crossings by large mammals and detected over 100 grizzly bear crossings per year. However, we felt that the true test of wildlife crossing structures was whether a whole population was using them. Were all these crossings by one bold male (as many skeptics believed) or was there equal mixing of use by males and females?

Clevenger and his graduate student Mike Sawaya did a DNA study to determine how grizzly bears were using the wildlife crossing structures. They snagged bear hair over three years, using barbed wire positioned at these structures.

Figure 1.3. Cougars using one of the Banff Wildlife Crossings Project's Bow Valley wildlife underpasses in Banff National Park, Alberta, across Highway 1. (Remote Camera Image, Anthony Clevenger.)

Figure 1.4. Grizzly bear with cubs using one of the Banff Wildlife Crossings Project's Bow Valley wildlife overpasses in Banff National Park, Alberta, across Highway 1. (Remote Camera Image, Banff National Park and Steve Michel.)

During each of the three years of this study, Sawaya found roughly a dozen bears using the crossings, with males and females equally distributed. This contrasts with Proctor's grizzly bear DNA findings on Highway 3 (which has half the traffic volume of Highway 1, but no wildlife crossing structures, and almost no grizzly bear crossings), but interestingly resembles Proctor's findings in the Flathead, where he found equal numbers of males and females crossing that roadless valley. Sawaya's study showed that the crossing structures on Highway 1 were working to create a more connected landscape for grizzly bears in Banff.[31] Additionally, this study provides a good example of how Clevenger and Sawaya's work independently built on and complemented work done by Proctor farther south, on Highway 3.

Clevenger has also learned about the importance of long-term monitoring. In seventeen years of monitoring, he found that it takes ungulates four to five years to adapt to using the structures, but for large carnivores this adaptation period is longer—from six to nine years. Monitoring projects that last only one to two years could fail to capture such success stories.

As of 2013, eleven large mammal species were using the Highway 1 crossings. This has resulted in an 80-percent reduction in mortality for all large mammal species, and an over 95-percent reduction for ungulates.[32]

Carnivores and Resilience

The concept of ecological resilience has great bearing on carnivore conservation. Ecologist C. S. ("Buzz") Holling first defined *resilience* as the ability of ecosystems to absorb disturbance and still persist in their basic structure and function.[33] With regard to the large carnivores (or any species), ecologists measure resilience by looking at how connected their populations are because of the importance of genetics to the health and vigor of populations, as we shall see.

Why is resilience so important? Climate change, widely evident along the Carnivore Way, will increase habitat fragmentation and shift plant-community ranges. Today, winters are measurably shorter than 30 years ago, with glaciers disappearing apace. Species that rely on snow and ice (e.g., the wolverine) may be hardest hit by climate change as they find their habitat shrinking to the highest peaks. According to Weaver,

> Nowadays, climate change has put such an exclamation mark on Holling's breakthrough concept about the importance of resilience. The question that we need to be asking is: How are these animals going to sustain their resiliency into a very different future? This brings the concept full circle to the idea that big, intact, well-connected landscapes offer the best opportunity for animals to move and find their new "normal" in terms of environmental conditions.[34]

To apply the concept of resilience, scientists begin by defining populations at three nested levels. (Animals like Pluie have demonstrated clearly the need to work at these multiple scales.) At the smallest scale we find the *individual* organism, and at the next scale the *local population,* also called a *deme.* At the largest scale lies the *metapopulation,* also known as a "population of populations."[35] Resilience accrues from these multiple levels: individual, local population, and metapopulation. However, specific mechanisms of resilience vary among species and among landscapes or ecosystems. Finally, scientists call the space separating local populations the *matrix.* Using grizzly bears in Banff as an example, a female grizzly is the individual. A female with cubs plus two or three other bears that inhabit the same small valley in the park make up the local population. All the bears in Banff historically formed one metapopulation, but Highway 1 and related human development has fractured it into two or more separate populations and diminished the natural movements and exchange of animals and their genes. To promote genetic connectivity and ensure that species don't become extinct (that is, that species show *persistence*), we need connectivity among local populations within a metapopulation.

Work by Weaver and his colleagues helps us understand how population ecology concepts such as local populations, metapopulations, the matrix, and genetics jointly affect the measure of resilience. Working at the three scales described above and using a number of factors, Weaver's team took a close look at the resilience of large carnivores in North America. They assessed individual resilience based on the capacity of an individual to switch between foods or habitats, should one of these factors decline. They evaluated local population resilience based on mortality rates, which include *stochastic* (random or unpredictable) events, such as climate fluctuations and disease. And at the metapopulation level, they evaluated resilience based on effective dispersal of young animals—which depends on the connectivity of travel corridors. They added two more dimensions to this resiliency framework: sensitivity to human disturbance and response to climate change. Taking all of the above into consideration, they found that some species, such as the wolf, are far more resilient than others, such as the wolverine, based on their natural history traits (e.g., reproductive rate and habitat needs).[36] We'll explore the concept of resilience for each species profiled in this book.

In Weaver's WCS *Safe Havens, Safe Passages* report, he explains that "in landscapes fragmented by human disturbance, successful dispersal is the mechanism by which declining populations are supplemented, genes are shared across the landscape, and functional connectivity of metapopulations is established." Further, he believes that the demographic dimension of connectivity (e.g., the ability of breeding females to cross Highway 3) may be equally or more important than the genetic dimension.[37] We can achieve connectivity via natural or human-made corridors positioned within the matrix, such as the Banff crossing structures. And to ensure persistence of a species, we need to ensure connectivity of metapopulations over very long distances and even across international boundaries, such as between the grizzly bears in Banff and those in Glacier. Only with adequate connectivity would these metapopulations have genetic diversity, because animals from these populations would have the ability to travel easily to find a mate and reproduce. Further, this sort of genetic diversity would make them more resilient.

Linking Landscapes and People

From the start, conservation biology has been about people and their values. Conserving and creating linked landscapes for large carnivores is, therefore, ultimately about our relationship with the natural world. To link landscapes, one must link people via collaborative conservation efforts. Heuer points out that "the challenge is going to be to stitch small-scale local efforts into something that's meaningful at a continental scale." In decades of collaborative Y2Y conservation, including a campaign to protect Alberta's Whaleback region, which contains abundant large carnivores, some of the oldest forest stands in Alberta, and land uses such as ranching, one of Francis' biggest personal lessons has to do with perseverance. "You never know when or how a success might occur. You just have to keep working at it."[38]

The large carnivores, such as Pluie the wolf, or the grizzly bears whose tracks Heuer found on his Y2Y trek, are emblematic of corridor ecology and more. Kevin Van Tighem, an environmental writer and former Banff National Park superintendent born and raised in rural Alberta, says that he has learned that large carnivore conservation

> enables us to get to the nub of things. You can focus on ungulates, you can focus on vegetation, and you can focus on natural processes like fire and flooding. They'll all give you insights, but carnivores take you to the very heart of the problem, because they're the hardest to live with. They're our competitors, potentially our predators, and from them run threads that reach into all those other things anyway.[39]

In chapter 2, we'll learn more about those threads—and how carnivores touch everything in the web of life.

CHAPTER TWO

The Ecological Role of Large Carnivores

Some animals become our teachers and guides—if we let them. Pluie the wolf (*Canis lupus*) was such an animal. She taught us that parks and protected areas, no matter how large, aren't big enough to meet all the needs of large carnivores. But with her return to places where wolves had existed, she and others of her kind also began to teach us essential lessons about why we need wolves to have healthy ecosystems. In her travels during the early 1990s, Pluie moved through much of the Crown of the Continent Ecosystem. At some points along her journey, she came very close to where my family and I live, although we had no way of knowing that then. But others like her returned to our valley at about the same time and taught us important life lessons.

In the late 1990s when wolves returned, we watched everything change. Shortly after my young daughters and I saw a pair of wolves course across the meadow on our land hunting a white-tailed deer (*Odocoileus virginianus*), we began to see changes in deer and elk (*Cervus elaphus*) behavior. With wolves around, these plant eaters could no longer afford to stand around brazenly mowing down shrubs and saplings. To avoid being eaten by wolves, they had to stay on high alert, taking quick bites and looking up frequently. As a result,

within five years of the wolves' return, the aspens (*Populus tremuloides*) began
to grow into the forest canopy, filling in the meadow and creating a home for
birds. The bold changes and relationships between species I witnessed made
me deeply curious and inspired me to study wolves as a scientist. This experience awakened in me a passion that I haven't lost, despite a decade of graduate
school and research: to try to understand the large carnivores and what they
need in order to thrive.

My research in the Crown of the Continent Ecosystem began in the mid-
2000s. Part of it entailed surveying wolf dens to learn about ecological relationships at these epicenters of pack activity. One of the most intriguing dens I surveyed was in Waterton Lakes National Park, Alberta. To better understand this
den's remarkable ecology, it helps to look at the trajectory of wolf presence here
and the development of our scientific understanding of the wolf's ecological
role.

Wolf history in the Crown of the Continent resembles what has happened
throughout the West. Seventeenth-century European settlers brought worldviews that called for dewilding the New World: systematically eliminating predators, fire, and forests. In the Crown of the Continent, humans killed wolves to
the point that, by the 1920s, there were none left.[1]

Over the next 50 years, in Alberta and northwest Montana, wolves began
slowly drifting down from the far north (the Yukon and Alaska), where they
hadn't been fully eliminated. These wolves followed the same pathway they had
used since the retreat of the ice sheets 12,000 years ago, moving along the cordillera formed by the Rocky Mountains. Indeed, their persistence in using this
corridor is what inspired me to call it the Carnivore Way.

Between 1940 and 1970, Waterton Lakes National Park records contain
many warden reports of wolf howls, tracks, and sightings.[2] Yet none of these
wolves survived long, due to human intolerance. We have a long history of killing what we fear. Rational or not, for centuries these fears kept wolf numbers
low to nonexistent, particularly in hunting and ranching communities. Predator removal (called "control") took place in national parks until 1934 in the
United States and until 1984 in Canada. The 1980s also brought a reduction of
indiscriminate predator control outside national parks. This meant that wolves

such as Pluie were able to travel far, often crossing provincial, state, and international boundaries. Thanks to our increased acceptance, by the mid-1990s wolves had successfully recolonized the Flathead and Belly River corridors and other remote valleys.[3]

Some places are more used by wolves as denning and hunting grounds than other places that seem to have equally good habitat. Nobody knows why that is. The Belly River Valley in Waterton became one of the places that drew wolves when they began to repopulate this area. Retired Waterton ecologist Rob Watt first documented them denning here in 1992.[4] Since then, the wolves had used the same den complex continually. This made the Belly River pack's den an important ecological study site that could provide insights about how wolves affect food webs. This site has drawn me as a scientist just as surely as it has drawn the generations of wolves that have denned here.

Since the 1920s, ecologists have understood that predator-prey relationships play an essential role in channeling energy flows within ecological communities. The presence of wolves touches everything in an ecosystem—from trees to songbirds to butterflies—because of how they influence their prey's behavior and density, and how that in turn affects the ways their prey eat and use a landscape.[5] This has to do with evolutionary relationships that have been in place for millennia. For example, a species such as a pronghorn antelope (*Antilocapra americana*) co-evolved with a species like the wolf chasing it. In his poem, "The Bloody Sire," Robinson Jeffers eloquently captured this relationship: "What but the wolf's tooth whittled so fine/ The fleet limbs of the antelope."[6]

These evolutionary relationships don't just pertain to wolves. They also apply to other predators, such as sharks (all species) and sea otters (*Enhydra lutris*), and in many other types of ecosystems. Worldwide, we're just beginning to understand the broad-scale ecological implications of predator removal—and return. Places actively being recolonized by wolves, such as Waterton and Glacier National Parks, or where wolves have been reintroduced, such as Yellowstone National Park, provide vital landscapes in which we can learn about these relationships.

As part of my studies, in late spring 2007, I surveyed the Belly River den with Rob Watt. Starting near the base of one of the peaks that surround this valley, we walked toward the den area on a narrow, overgrown game trail. Young aspens, straight and unbrowsed, grew closely along both sides of the trail, the understory shrubs lush and wet from late spring rains. We quickly saw evidence that we were traveling in the right direction—wolf scats deposited tellingly every 100 feet or so as territorial markers. Occasionally we found enormous, fresh grizzly bear (*Ursus arctos*) scats. Over the rush of wind through conifers, we heard the unmistakable braid of wolf voices, and around the next bend in the trail, we heard something big crashing through the woods. Hooting and hollering to let the bears know we were there, we cut upslope. The trail narrowed to nearly nothing at times, but we managed to follow it.

At a slender creek, we found wolf tracks in the streambank mud. Just past the stream, Watt led me off the game trail and we were engulfed by what rural Albertans call the "shintangle"—a dense thicket, here composed of snowberry (*Amelanchier alnifolia*), serviceberry (*Symphoricarpos occidentalis*), rose (*Rosa* spp.), and willow (*Salix* spp.). We passed aspens cut by beaver (*Castor canadensis*), the stumps weathered silver. After a few miles of bushwhacking, we descended into the swale of a wet meadow ringed by pale green aspens and inky conifers. And on the meadow's rim, we saw dark hole after dark hole dug into the sloping earth—a series of wolf dens necklacing the forest edge. We weren't there to count dens, but to determine whether this pack had produced pups.

We soon found what we were looking for. In a grassy clearing, we saw a partially chewed, bloody deer skull, part of the hide still attached. All at once, in a shaft of sunlight in another grassy opening, we stumbled upon three small, gray-colored wolf pups playing tug-of-war with a scrap of deer hide. They became aware of us at the same moment we became aware of them, and they instantly fled into the woods. We stood there, too stunned to say much. We had our answer and left quickly, to avoid disturbing the wolves further.

Afterward, Rob and I talked about returning to this den to measure its ecology later in the season, after the wolves were gone from the den. When pups reach the age of eight to ten weeks, packs typically move them to more distant places within their territory, called *rendezvous sites,* to begin to teach them how to behave like wolves. We thought it would be interesting to come back and measure the potential mark of wolf presence on the ecology of this den—food web relationships called *trophic cascades.*

At their most fundamental level, trophic cascades are relationships. The word *trophic* means "related to food." In the system that I studied, trophic cascades involved relationships among wolves, their primary prey (elk, in this case), and the plants that nourished elk (aspens).[7] As in a waterfall, in a trophic cascade predators exert a controlling influence on prey abundance (called *density-mediated effects*) and prey behavior (called *behaviorally mediated effects*) at the next lower level, and so forth through the food web. Remove an apex predator, such as the wolf, and elk and other ungulates grow abundant and fearless, damaging their habitat by consuming vegetation unsustainably (fig. 2.1). As elk literally eat themselves out of house and home, biodiversity plummets. Return wolves to ecosystems and elk become more wary, spending more time with their heads up above their shoulders (called *vigilance*), staying on the move, consuming vegetation with more restraint (fig. 2.2). Ecologists Joel Brown and John Laundré called these behavioral relationships between predators and their prey *the ecology of fear.*[8]

Figure 2.1. Lamar Valley elk eating aspens in Yellowstone National Park, Wyoming. (Photo by Cristina Eisenberg.)

Figure 2.2. Vigilant Lamar Valley elk in Yellowstone National Park, Wyoming. (Photo by Steve Clevidence.)

Ideas about the link between predators and their prey aren't new. In the 1920s, wildlife ecologist Olaus Murie surveyed ungulate habitat in the western United States. On visits to elk winter ranges (low-elevation grasslands that don't have deep snow in winter) where wolves had been removed, he found no aspens growing above elk browse height (seven feet). A highly nutritious tree species, aspen provides essential food for elk in winter when grasses have been depleted. Twenty years later, Aldo Leopold conducted a continental-scale study of the effects of wolf absence on ungulates. From Quebec to Mexico, he found that wherever wolves had been removed by humans, ungulates were wiping out their food sources, especially woody species such as aspen and balsam fir (*Abies balsamea*).[9]

Trophic cascades are based on the Green Earth Hypothesis, a provocative idea put forth in 1960 by Nelson Hairston, Frederick Smith, and Lawrence Slobodkin (collectively known as HSS). They proposed that how herbivores (plant eaters) eat determines how vegetation grows. They further suggested that predation shapes these patterns and that, unchecked by predators, herbivores will have great impact on vegetation. HSS concluded that as long as there were predators, the world would be green, because predators affect their prey, and that in turn influences plant ecology from the top down.[10]

Not All Species Are Equal

All species matter, but some have more powerful ecological effects than others. In the 1920s, British ecologist Charles Elton created the food pyramid concept, with apex predators at the top, herbivores at the next lower level, and plants at the bottom. He suggested that apex predators exert a strong influence on food webs from the top down, working together with solar energy, which makes plants grow, to create healthy ecosystems. Elton also noted that apex predators, much less abundant than their prey, are far more vulnerable to extinction.[11]

Thirty-five years later, Elton's ideas inspired ecologist Robert Paine's work in Washington's rocky intertidal zone. To test the importance of apex predators, he removed the sea star *Pisaster ochraceous*, a carnivore that rapaciously preys on the mussel *Mytilus californianus*. *Pisaster* consumes mussels by extruding its stomach, engulfing its prey, and then liquefying it with digestive enzymes. Once liquefied, *Pisaster* absorbs its food and pulls its stomach back inside its body.

After weeks of systematically hurling sea stars into the ocean in his promethean experiment, Paine had rewarding results. Where *Pisaster* flourished, so did the vegetation. Where he'd removed *Pisaster*, the mussels took over, crowding out other species and consuming most vegetation. Biodiversity dropped sharply in mussel-dominated sites because mussels, by eating plants, eliminated habitat for other species.

Paine used the interesting relationships he observed between sea stars and mussels as the basis for his *keystone species* concept. He envisioned a keystone species as a dominant predator that consumes and controls the abundance of a particular prey species, an herbivore. This prey species competes against other herbivores. In Paine's world, a keystone species fits into an ecosystem like the keystone of a Roman arch. His metaphorical term suggests that when you remove the keystone, ecosystems—like arches—will fall apart.[12]

In 2005, Michael Soulé and his colleagues coined the term *strongly interacting species* to expand on the keystone concept. They defined strongly interacting species as those whose presence, even in small numbers, have disproportionately large effects on food webs. While strongly interacting species could be carnivores, they could also be herbivores such as bison (*Bison bison*), beavers, and prairie dogs (*Cynomys ludovicianus*).[13] Collectively, Paine's and Soulé's ideas demonstrated that everything in nature is connected, and if you tug on one species, you will affect others—sometimes with far-reaching consequences.

The Mark of the Wolf's Tooth

I returned to the Belly River pack's den in August 2007 to find out whether areas of especially intense wolf use would show the strongest trophic cascades. In other words, if Paine's keystone species concept applied to my research site, then wolf presence would mean wary elk, and I'd expect to find thriving young aspens growing above elk browse height.

My field crew and I entered the meadow and walked along its edge, examining the dens until we found the one most recently used. Set into the root ball of a spruce (*Picea* spp.), the massive den had four cavities, each one smooth, deep, and clean, with layers of wolf tracks pressed into the bare earth within.[14] Pup chew toys—elk jawbones, a beaver skull, and a purloined human shoe—lay strewn about nearby.

We proceeded to measure the food web at this den. This meant documenting wolf, elk, and other ungulate activity, as well as aspen demographics (how the aspens were growing), within a 300-yard circumference of the den. Radio-collar data on my project suggested that the intensity of wolf use was highest within this radius. In our sampling, we systematically counted wolf scats, all the piles of deer, elk, and moose (*Alces alces*) pellets, as well as carcasses. We measured sapling height and whether the aspens had been browsed, as evidenced by chisel-like bite marks. We counted aspen sprout density and looked at aspen demographics—how many stems had grown above browse height (called *recruitment*), and the relative ages of these trees. Additionally, we stratified our sample into four quadrants oriented toward each cardinal direction. This would enable us to determine whether differences in aspen growth and recruitment into the forest canopy might be due to aspect (i.e., warmer, south-facing slopes versus cooler, north-facing slopes). Over the next week, we conducted leave-no-trace sampling (taking measurements and nothing else), working rapidly to minimize our presence and impact at this sensitive site.

Later, when I analyzed my data, I found the mark of the wolf's tooth inscribed indelibly in the pronounced ecological patterns here. I found no piles of elk or deer pellets within a 300-yard radius of the den, and just five piles of moose pellets. Within the same radius, I found 186 wolf scats and 57 carcasses. Only 8 percent of the aspens showed evidence of ungulate browsing. The juvenile aspens were growing tall, beyond elk browse height. This release from herbivory had created a classic trophic cascade signature here, called a *recruitment gap* in an aspen community: lots of old aspens, no middle-aged aspens, and a flush of young aspens rapidly growing into the canopy.

Park records indicated that, before wolves returned, the Belly River Valley had been heavily used by elk as part of their winter range. A herd of 200–500 elk would come here in winter, which for an elk in this extreme landscape begins in October and ends in late May.[15] The elk would stay sheltered in this valley, eating grass. When the grass was depleted or buried by snow, they'd eat tender, nutrient-packed aspen shoots. In my den survey, I didn't find evidence of any aspens growing into the forest canopy during the seven decades when wolves had been absent from this valley.

Five years later, in the summer of 2012, I revisited the Belly River Valley and found a completely different story. Initially, this visit was just part of my routine annual survey to determine whether the wolves were still denning here. That year the wolves had been under considerable pressure from ranchers outside the park. Unsurprisingly, I found that the wolves were no longer using the den complex. I didn't see any scats, tracks, or carcasses more recent than two years.

Troubled as a conservationist and intrigued as a scientist, I remeasured the ecology of this site, using exactly the same methods I'd used in 2007. One week later, when I analyzed my data, I found the opposite of what I'd found five years earlier. This time, there were 98 piles of elk pellets and 6 piles of moose pellets. I only found one wolf scat and twenty carcasses. The aspens bore the blatant signature of too many elk and not enough wolves. Sixty-three percent of them had been browsed. Some had been browsed at ground height—mowed down by elk.

I considered what else could have influenced this shift in how the aspens were growing at the den. The climate hadn't differed appreciably between 2007 and 2012, the years when I collected the data. The elk population had remained

the same overall in the park. Yet the evidence before me told the incontrovert-
ible story of the strong behavioral effect of keystone predators on whole food
webs. When wolves were using this den, elk avoided the site and browsing in the
area was low, enabling aspens to thrive. When wolves stopped using this den,
elk returned, heavily browsing the aspens. Yet bold as this signature appears
here, like all things in nature, these relationships among predators, prey, and
plants aren't simple.

Top-Down vs. Bottom-Up: Nature's Tangled Bank

In the mid-1960s, ecologist William Murdock suggested that food (bottom-
up control) has a stronger influence than predators. In his famous rebuttal to
HSS's ideas, he claimed that the world may be green because not all plants are
palatable to herbivores. He argued that the sun's energy, plant growth, and dis-
turbances such as fire move nutrients through ecosystems from the bottom up
(e.g., causing fire-adapted species such as aspens to grow more energetically).
He further asserted that predators are unnecessary to keep the world green.[16]

HSS's and Murdoch's two opposing views of nature sparked a heated argument about whether ecosystems are regulated from the top down or from
the bottom up. This argument continues to split the scientific community today. Some consider it a rather ridiculous debate, because the natural world is a
system that contains multiple pathways and relationships and types of energy.
However, science often lurches forward dialectically, perhaps because human
nature compels us to find simple and unique explanations to complex problems. Because of my own research with wolves, elk, and aspens, I became curious about top-down versus bottom-up effects. Some of my study areas had had
fire in them—a bottom-up effect. Ecologists have long identified fire as a strong
bottom-up influence that stimulates plant growth. Figure 2.3 depicts the flush
of aspen growth after a fire in Glacier National Park.[17]

While trophic cascades science in aspen communities is still relatively new,
a variety of fascinating studies have been done in this field over the past decade
and a half. Most have been *observational*, which means that researchers simply documented what they found. *Experimental* work means manipulating the

Figure 2.3. Flush of aspen growth after a fire in the North Fork area in Glacier National Park, Montana. (Photo by Brent Steiner.)

elements one is studying (such as elk numbers or aspen stands, in this case), a more rigorous way of testing a hypothesis. Given the newness and observational nature of trophic cascades science in aspen communities, the knowledge gaps in this field are still huge. For example, we've learned that wolves are linked to strong direct and indirect effects, which cascade down through multiple trophic levels. However, we're just starting to measure how context can influence these effects.William Ripple and his colleagues conducted the earliest trophic cascades research in Yellowstone National Park. In this groundbreaking observational study, they wanted to determine how reintroduced wolves were influencing elk browsing patterns and aspen recruitment through trophic cascades mechanisms. Subsequent wolf trophic cascades studies in Yellowstone and elsewhere built on their study.

In the 1800s, Darwin presciently described nature as a "tangled bank." This complexity results from myriad species and their relationships with other species and all the things that can possibly affect them individually and collectively, such as moisture, disease, and disturbance.[18] Science works incrementally, taking us ever deeper into nature's tangled bank as we investigate ecological questions. Each study answers some questions and begets new ones; each study leads to another as we learn new things. Sometimes we find contradictory results. Learning how nature works requires what Leopold called "deep-digging research" in which we keep searching for answers amid the clues nature gives us, such as the bitten-off stem of an aspen next to a stream where there are no wolves.[19]

Ripple and his colleagues used a variety of observational methods. To measure elk use of their study area, they used elk pellet groups; to measure wolf use, they used radio-collar data. To illustrate differences in the ecology of high and low wolf-use areas, they used aspen-sprout heights and the percentage of browsed young aspens. They determined that streams had high wolf use and greater predation risk for elk. Along these streams, they found fewer elk pellets and taller aspen sprouts, which they interpreted as evidence of a wolf-driven trophic cascade. They found no significant effect of fire on aspen density or height. One of the most important aspects of this study is that Ripple and colleagues set up permanent plots, which they've continued to monitor regularly since 1999. As such, this study provides valuable baseline information about potential wolf effects on elk and aspen in Yellowstone.[20] The longevity of this study and the fact that it began within three years of the wolf reintroduction give it significant historical importance, as it is chronicling changes in this wolf–elk–aspen system.

About a dozen subsequent observational studies in Yellowstone and beyond looked at the effect of wolf presence/absence on aspen communities. All hypothesized that behaviorally mediated effects (e.g., wolves scaring elk) would be linked to aspen recruitment. All found top-down effects, indicated by aspen recruitment above browse height. Further, most of these studies considered both top-down and bottom-up factors (e.g., fire, climate, soil moisture) as well as their interaction.[21]

Looking at trophic cascades outside of aspen communities, James Estes and other scientists have discovered top-down keystone predator effects in a variety of ecosystems worldwide. During the 1980s and 1990s, dozens of experimental studies in aquatic and terrestrial invertebrate systems identified strong causational links between predators and trophic cascades. I profiled many of these studies in my book *The Wolf's Tooth: Keystone Predators, Trophic Cascades, and Biodiversity.* However, scientists have examined the theory behind trophic cascades and found rich, complex food web relationships. The context and strengths of these effects remain to be fully explored in terrestrial ecosystems, in part due to the difficulty of experimenting with large mammals, which can't be manipulated ethically (e.g., by removing them, like sea stars were removed in Paine's study of mussel beds).[22]

Some recent studies have found bottom-up effects as well as what are called *trophic trickles*—situations in which a keystone predator may be present, but has only weak, indirect effects on vegetation. In Banff National Park, Mark Hebblewhite and his colleagues looked at whether variations in wolf presence would correlate positively to the strength of top-down effects. In this observational study, they measured wolf presence using radio-collar data and included a site from which wolves had been partially excluded (the Banff town site). They found no significant effects of wolf presence on aspen recruitment into the forest canopy. However, they found a relationship between fire and aspen recruitment. Sites that had had fire within the past 25 years showed more aspen recruitment than sites that hadn't had fire. And in willow communities, they found a significant release from herbivory that correlated positively to wolf presence. This led them to conclude that while wolves exert powerful top-down effects in willow communities, in aspen, bottom-up effects (disturbance) interact with top-down effects (wolf presence).[23]

In Yellowstone as well, some researchers have found trophic trickles. Matthew Kauffman and his colleagues did an experiment to look at behaviorally mediated trophic cascades on the Northern Range. They randomly placed small ungulate exclosures (wire-mesh cages) over aspens in areas of high, moderate, and low predation risk. To measure predation risk, they used radio collars on wolves in order to identify the sites where wolves were killing elk. High predation risk sites included riparian areas. After measuring hundreds of aspen plots, they found that these saplings weren't recruiting outside exclosures, regardless of predation risk, and that elk browsing didn't decrease in areas of higher predation risk. Therefore, they concluded that wolves weren't modifying elk browsing behavior in Yellowstone.[24]

Inspired by the changes I had witnessed on our land after wolves returned, I decided to study these effects in the Crown of the Continent. Because of its low human presence in winter, when elk are on their winter range, Waterton-Glacier International Peace Park was perfect for my studies. From 2007 to 2013, I conducted landscape-scale observational research in which I looked at relationships between wolf predation, elk herbivory, aspen recruitment, and fire in three valleys. My study areas were the North Fork Valley in the western portion of Glacier National Park; the Waterton Valley in Waterton Lakes National Park; and the Saint Mary Valley in the eastern portion of Glacier. All valleys lie in elk winter range, but have three different wolf population levels (Saint Mary: low; Waterton: moderate; North Fork: high), which represent three corresponding levels of long-term predation risk (the probability of an elk encountering a wolf). All three valleys have a very high elk density, among the highest recorded in North America (28–75 per square mile), regardless of wolf presence. Ecological characteristics (e.g., climate, soils, elevation, plant communities) are the same among all three valleys. Ninety percent of the North Fork had burned in the previous twenty years. My study plots in aspen in the other valleys hadn't burned since 1860.

The Belly River den lies in the Belly River Valley, in Waterton. I didn't include it in my comparison of three valleys, because of its different ecological characteristics. It lies in the montane zone, at a higher elevation than the other valleys, and during the course of my doctoral research it wasn't used by elk for winter range (although it was used until the early 1990s). Accordingly, I studied trophic cascades in the Belly River separately, as a den-site survey.

Comparing the three valleys, I wanted to know if an area that had lots of wolves present all the time (the North Fork Valley) would have stronger trophic cascades than an area that had wolves occasionally passing through (the Saint Mary Valley), and how fire would influence these effects. I also wanted to know if, in a valley that had lighter wolf presence than the North Fork (Waterton), wolves would drive trophic cascades.

Predictably, I found less browsing on aspens in the North Fork, where there was a high wolf population, suggesting a top-down effect. However, when I looked more closely at these relationships, I was astonished to find that in the

North Fork, the aspens were only recruiting above browse height where there'd been fire. In unburned sites, the elk were still keeping the aspens trimmed low. My findings suggest that in this ecosystem wolves need the bottom-up effect of fire, which makes aspens sprout more vigorously, in order to drive trophic cascades. I didn't find trophic cascades in my Waterton or Saint Mary study sites (which hadn't burned since the mid-1800s), regardless of wolf presence.[25] This suggests that along with wolves, fire may be considered another "keystone" force of nature. I also looked at whether wolves would have behaviorally mediated effects on elk, by scaring them. Again, some of my findings were unsurprising. For example, in Saint Mary, which lacked a wolf population, none of the predation risk factors I examined (e.g., obstacles to escaping or detecting a wolf) had any effect on elk behavior, and the elk were relatively complacent. Elk in the other two valleys were more wary, but their strategies to avoid being preyed upon by wolves differed markedly. In Waterton, wolf activity and presence varied, due to humans legally killing wolves outside the park. The elk in this park didn't know where wolves were or when they'd turn up next. Consequently, Waterton elk made large groups, because there's safety in numbers. However, in the North Fork, where wolves were everywhere in winter range, elk made smaller groups, because wolves have a harder time detecting small herds of elk.[26] In sum, I found that the risk of wolf predation alone didn't control the food web relationships I observed. Bottom-up forces (fire) and top-down forces (wolf predation) worked together in valleys that had well-established wolf populations.

In my Belly River den study, I found very strong wolf effects. When wolves were present, the aspens thrived, growing rapidly above the reach of elk, providing habitat for songbirds and a wealth of other species. When wolves were removed by humans outside the park, elk use of the site exploded, preventing the aspens from growing above the reach of elk and into the forest canopy. Further, this site hadn't burned since 1860, so none of what I observed there was affected by fire.

The differences in the results of my two studies have to do with scale. On the relatively small scale of a wolf-den survey (a 300-yard radius), you can find very site-specific wolf influences. But when working on a larger scale (the

4,000-square-mile area of the three valleys), you find greater variation in these effects, influenced by context. This is because the larger your study area, the more *heterogeneous*, or diverse in character and content, your data will be. For example, if you measure 20,000 aspens in three valleys, you may find more diversity than if you measure 500 aspens around a wolf den. The greater diversity in the larger-scale study happens because, as you cast your net over a broader area, you pick up more stochastic effects (e.g., aspens blown down by the wind and aspens that have disease) and other sources of variation.

Early Yellowstone studies took necessary first steps to assess wolf-driven trophic cascades. As with any science, subsequent research delved deeper, building on earlier findings and uncovering both top-down and bottom-up effects. This makes sense. Today we're learning that in the tangled bank that we're studying, top-down and bottom-up forces intertwine in a trophic dance between wolves, elk, and aspens that has been going on for millennia. Our inquiries sometimes lead us to apparently contradictory findings. But given the astonishing complexity of our world, such findings reveal fascinating facets of nature and add to our growing picture of food web relationships.

When Is a Carnivore a Keystone? (and Other Burning Questions)

Scientists such as John Weaver depict the Crown of the Continent as containing one of the highest densities of large carnivores south of Alaska and the Yukon. If so, would other species besides the wolf drive trophic cascades? And if not, why not? For example, I found that in my low-wolf study area, even with grizzly bears and cougars (*Puma concolor*) present in high numbers, elk continued to wipe out aspens. Why was that? As with most ecological questions, the answer has to do with context. Or, as I often tell my students in response to their questions about ecological relationships, "it depends."

The cougar has context-dependent food web effects. This solitary carnivore mainly kills and eats deer. After killing a deer, a lone cougar can only consume part of it because of the large amount of meat the carcass provides. It typically eats as much as it can, buries the rest with dirt, twigs, and leaves, and takes a nap. While the cougar sleeps, a bear will often find and steal a cached carcass.

Other animals, such as coyotes (*Canis latrans*), also feast on this stolen bounty. In terms of trophic cascades, in places where the most abundant herbivore is the deer, some studies have found that cougar presence reduces deer browsing on woody plant species. However, cougar-driven trophic cascades don't exist where elk are the most plentiful herbivore, because cougars don't kill many adult elk.[27] However, cougars do kill elk calves seasonally, which can affect elk populations by reducing the number of elk that grow into adults.

Like the cougar, the grizzly bear doesn't drive top-down trophic cascades in most places. A grizzly bear is an omnivore, which means it eats both vegetation and meat. It disperses the seeds of the plants that it eats via its scat, thereby helping shape plant-community distribution. Also like the cougar, it usually doesn't prey on adult elk, instead killing elk calves. And in places like Yellowstone, the grizzly spends considerable time scavenging wolf kills, forcing wolves to kill more often in order to nourish themselves. Additionally, the fact that the grizzly hibernates during part of the year tends to weaken its food web effects. All of these factors make quantifying the grizzly bear's food web role challenging.

The lynx's (*Lynx canadensis*) food web role is simpler and closely resembles the wolf's. A solitary carnivore, the lynx preys primarily on snowshoe hares (*Lepus americanus*). Hares have nine- to eleven-year population cycles. As snowshoe hare numbers rise, lynx numbers rise, and as snowshoe hares decline, lynx decline. Lynx predation on snowshoe hares has powerful food web effects because it reduces prey available for raptors, causing raptors to decline. Lynx predation can also indirectly affect vegetation. Denali National Park bears the unmistakable marks of a snowshoe hare population high combined with a lynx population low: miles and miles of willows with dead limbs. In the 2000s, Denali's hare population skyrocketed. In the two years that it took for lynx to catch up, these herbivores feasted on willows, girdling them by removing the bark all the way around stout limbs. Since willows send up multiple shoots, girdling doesn't kill the entire shrub, just some of its branches. Willows provide important habitat for songbirds such as willow flycatcher (*Empidonax traillii*), Wilson's warbler (*Wilsonia pusilla*), and yellow warbler (*Dendroica petechia*). The bare, dead willow branches created during a hare high and lynx low can negatively affect these songbird species and others, creating a "truncated" trophic cascade.[28]

We have only a tenuous understanding of the ecological role of the other large carnivores in the West. For example, the wolverine (*Gulo gulo*) is both a scavenger and a predator. Its scavenging behavior may limit its ability to exert top-down effects in ecosystems. However, the relative paucity of wolverine research, combined with this species' scarcity, leaves us with an incomplete picture of its ecological role. Similarly, we know little about how jaguars (*Panthera onca*) interact with whole food webs, because this species is so elusive and mysterious. Yet, given what we do know about large carnivores, we can safely say that the wolverine and jaguar both function as strongly interactive species, and in some circumstances may have a keystone role.

Regardless of large carnivore species' precise role in a food web, we've learned that trying to restore an ecosystem without restoring the large carnivores resembles trying to fix an engine without using all its parts. In the 1940s, Aldo Leopold wrote, "To keep every cog and wheel is the first precaution of intelligent tinkering."[29] Sage advice. However, it isn't enough to have large carnivores present. For them to fulfill their ecological roles, they need to be present in sufficient numbers.

Today we use the concept of *ecological effectiveness* to evaluate whether the population density and activity of a keystone species in a specific area is sufficient. In this sense, an ecologically effective population is one capable of stimulating top-down trophic cascades. Like everything else in ecology, ecological effectiveness depends on context. We've seen how, in the case of the wolf, top-down (predator-driven) forces in large-mammal systems may be assisted by bottom-up (natural-resources-driven) energy created when events that kill forest stands (e.g., fire, wind storms) happen. In such cases, a lower number of wolves could trigger top-down trophic cascades because bottom-up energy created by fire is enabling vegetation to grow more rapidly and vigorously than usual.[30]

Even as we strive to expand our knowledge of species interactions, biodiversity is decreasing at an alarming rate. Extinction is forever, but it's not new to this planet. There are two types of extinction: *background extinction*, which is an ongoing natural process, and *sudden extinction*, which is caused by catastrophic events such as large asteroids hitting Earth. Background extinction makes room for new species and typically occurs at the rate of one to five species per year.

However, today we may be losing as many as 30,000 species per year—a sobering three species per hour, and an average extinction rate of 6 percent per decade. Nobody really knows exactly when this increased rate of extinction began—likely sometime during the Anthropocene. However, scientists first became aware of it in the early 1980s. Further, the rate of extinction is at best an estimate, because we have not yet identified all species on Earth. Five major extinction waves have passed over the earth in the last 440 million years. We're currently in the sixth such extinction wave, this one the result of the human destruction of ecosystems in the Anthropocene epoch—the Age of Man.[31] Large carnivores can help slow extinction because they increase biodiversity, thereby creating more resilient ecosystems. And an ecosystem that contains more species can better withstand change.

Given the fact of ongoing extinction, arguing about what exactly carnivores do ecologically and why we need them is like fiddling while Rome burns. Large, meat-eating animals improve the health of plant communities and provide food web subsidies for many species. There are things we don't know and disagreements about what we do know. But given the indisputable fact of human-caused extinctions in the Anthropocene, a precautionary approach to creating healthier, more-resilient ecosystems means conserving large carnivores.

Into the Far North: Boreal Ecology and Trophic Cascades

In July 2012, I drove from Montana to Alaska to learn more about carnivores in the northern portion of the Carnivore Way. My traveling companions included my husband Steve, and our niece and nephew Alisa and Andrew Acosta. My friend Donna, a science writer who hails from Kananaskis Country and has a keen interest in botany, would be joining us in Alaska.

Our journey took us north of Banff, into Jasper National Park and the wild heart of the Canadian Rockies. Dramatic scenery unfurled along the way: gunmetal-hued rock pinnacles, snowfields, wild rivers, and thick coniferous forests. Established in 1907, Jasper is 2.7 million acres in size and contiguous to Banff. It represents an important portion of the connected wild landscape that extends from Glacier National Park to Alaska.

As we approach Jasper, I note that the tall, dark spires of spruces dominate the thick forest. We've entered the boreal (that is, northern) biome. In North America, the boreal biome covers almost 60 percent of Canada's land area and spans the continent from Newfoundland to Alaska and the Yukon. Life-forms here have adapted to cold, snowy winters and short, hot summers.

The southern portion of this biome contains boreal forest, such as the one we're traversing. Boreal ecologists William Pruitt and Leonid M. Baskin refer to this far northern forest as a green nimbus around the northern end of our planet. Primarily coniferous, boreal forest consists of spruce interspersed with wetlands that include bogs and fens. Dominant tree species are black spruce (*Picea mariana*), white spruce (*P. glauca*), larch (*Larix laricina*), and lodgepole pine (*Pinus contorta*). Because of the short growing season, generally infertile, shallow soils, and poor water drainage, boreal forests grow slowly.[32]

Permafrost, or permanently frozen ground, shapes the ecology of the northern portion of the boreal biome. Defined as soil, sediment, or rock that remains at or below freezing for at least two years, permafrost mostly occurs in North America in Alaska, the Yukon, and northern Canada, between the latitudes of 60° and 68° North. Jasper, at 53° N, lies well below the permafrost zone.[33] The permafrost zone contains sparse, stunted forest, called *taiga*. Soon enough on our journey, we'll see how permafrost affects prey availability for the large carnivores—and trophic cascades.

The Mother of All Transects

From Jasper we travel west to Prince George, on Highway 16, also known as the Yellowhead Highway, and from there north to Dawson Creek, British Columbia, where we'll take the Alaska Highway 1,400 miles to Delta Junction, Alaska. This notoriously gnarly highway has only recently been paved in its entirety. Along the way, we'll cross several ecological provinces: from boreal forest to inland rainforest and then north into permafrost taiga and tundra—treeless frozen ground that supports only low-lying vegetation.

As we leave Prince George, I have the sobering realization that the Internet mapping tool I used to determine the mileage from Jasper to Dawson Creek was wrong—by about 300 miles. Not a small thing. Our itinerary requires that we arrive in Anchorage, Alaska, within three days, so that Steve can fly back home and we can pick up Donna, who'll be joining us there. At the same time, as I become aware of the extra miles, I realize that on this journey what we actually are doing is pulling the mother of all transects, figuratively speaking. This second realization is one of the defining moments of this trip.

A transect is an imaginary line, often marked with a measuring tape. Ecologists gather data along that line, typically in plots. We use those data to test hypotheses or to answer research questions, such as how dense the forest is and what factors might be influencing that density. When finished gathering data, we simply pull up the tape, thereby leaving no trace—the sort of sampling we did at the Belly River den.

While I didn't set out to pull a transect from Waterton to Alaska, as an ecologist, I can't help but think this way. So visualize, if you will, our route from Waterton to Alaska as a continental-scale transect. Along this transect I'm assessing apparent trophic cascades. In it I note predator and ungulate presence and plant-community characteristics. Aspens and willows provide important food, so their condition can tell me much about the rest of the food web.

A rapid roadside assessment of trophic cascades patterns (e.g., vegetation releases from herbivory) isn't as far-fetched as it may seem. Hydrologist Robert Beschta developed such methods to evaluate forest dynamics in Oregon riparian cottonwood (*Populus* spp.) communities. Done from a distance, it involves quickly categorizing focal species, such as aspen, by age class (e.g., mature trees, juveniles, and sprouts).[34]

Slightly north of Dawson Creek and 600 miles south of the British Columbia-Yukon border, we spot a lone elk—the last one we'll see on our trip north. Elk occur in very low density in the southeastern Yukon near the border, most likely moving up from British Columbia. In southern British Columbia and Alberta, and in the US Rocky Mountains, elk are the dominant large herbivore, present in much greater density and biomass than deer or moose. As we travel north, elk decline in number, moose and caribou (*Ranger tarandus*) increase, and deer remain low.

I'm so fascinated by the shifting patterns I'm observing, and others that I know are here, that I let Alisa and Andrew take over driving. This frees me to focus fully on plant and animal communities and to take field notes and photographs. To make up those extra 300 road miles, we drive into the night. At around 11:00 p.m. we notice that the sun isn't setting. It hangs midway down in the sky, like a glowing red rubber ball, bathing the landscape in ethereal, golden half-light. At this latitude, 60° North, we've reached the permafrost zone, where midsummer days are twenty hours long, and midwinter nights similarly long. We begin to see subarctic tundra. This frozen soil supports only low-lying vegetation.

I read the landscape flashing by and ask my traveling companions to find me some trophic cascades. To help pass the miles, I introduce Alisa and Andrew to terms like *ecotone*—the place where two habitat types meet, such as the ecotone between forest and grassland, in which we spot many moose and the aspens appear to be releasing (i.e., growing vigorously into the forest canopy). A trophic cascade! Well, maybe. We talk about the many bottom-up factors that could be influencing such a release, such as soil type, moisture, aspect, and root stimulation from disturbance (e.g., the road cut). These factors can cause some tree species, such as aspen, to sprout and grow more energetically. Then there's also the effect of traffic, which changes wildlife behavior, as we've seen in Banff.

More than a little giddy from all this drive-by ecology on such little sleep, I realize that we're on a marathon continental-scale ecology fieldtrip, one that will forever change the way I look at nature. Along the way I learn that Soulé was right, carnivore conservation is a matter of scale—the bigger the better.

When we cross from British Columbia into the Yukon and the permafrost zone, we notice other shifts in ungulate distribution. Caribou occur in low numbers up to Jasper. But then their density increases significantly in central British Columbia, and is highest the farther we go north. Moose dominate the southern Yukon. And as we move north, caribou rapidly increase, becoming the dominant Yukon herbivore. Specifically, 300 elk, 70,000 moose, and 220,000 caribou live in the Yukon. Sheep (*Ovis* spp.) and mountain goats (*Oreamnus americanus*) are scarce, but their presence and ecological importance will grow once we enter Alaska.

As the human population decreases, the carnivore population increases. From Prince George, British Columbia (pop. 72,000) to the north, the human population is very low. For the next 2,000 miles on our route between there and Anchorage, Alaska (pop. 290,000), there is only one city with more than 5,000 people: Whitehorse, Yukon (pop 27,000). People who travel this road call it an "outlaw" highway, due to its roughness, wildness, and crime. This north country has a thriving carnivore population, even though the federal and provincial government gives them little protection. Wildlife managers don't know exactly how many carnivores live here because they don't spend as much time counting them as they do game species such as moose and caribou, but they know there are many. Carnivore numbers in the areas we're traversing far surpass what we have south of the US-Canada border. All this makes me mighty curious about the trophic cascades I might find here.

As the road miles accumulate, I begin to see that shifts in ungulate-species composition have profound implications on the ecology of the plant communities along the way. This has to do with what ungulates eat. Elk mainly eat grass and aspen. Consequently, where elk dominate, I find aspens over-browsed, with little recruitment regardless of the carnivore population. Even in the Canadian national parks, I find patchy trophic cascades in aspen. Moose primarily eat sedges and riparian vegetation and willows. Where they dominate, aspens are fine, with many young trees growing into the forest canopy. Better than fine, they're some of the most splendid-looking aspens I've ever seen, growing in lush thickets. Moose impacts can best be evaluated via willows, but most willows I see are eight to twelve feet tall, not showing evidence of being suppressed by moose.[35] Northern caribou mostly eat lichen, so they have little herbivory impact on aspen forests. Spruce, the dominant tree species throughout the boreal zone, aren't palatable to ungulates and provide little nutrition to them.

These patterns provoke further wonder about ecological relationships. What is the role of disturbance (fire) and other factors in driving aspen recruitment? The region we're traversing has had patchy fire, but Canada has a fire-suppression policy beyond its national parks. Are wolves present in ecologically effective numbers in places where they can be freely hunted and trapped? And what about the low net primary productivity (NPP) of the far north? (NPP refers to energy flow in ecosystems, a measure of biomass, or ability of things to grow, driven by sunlight, moisture, and photosynthesis. As one approaches Earth's poles, NPP drops significantly, as does the number of species that these extreme ecosystems can support.[36]) And what about the road effect? What if I were able to randomly position this mega-transect far from this highway? How might this affect observed trophic cascades? Lots to ponder as the road miles pass.

We approach the Yukon-Alaska border the next day, not having slept much in the previous 36 hours. Unfiltered impressions fill my sleep-deprived brain: of how much more open and intact this landscape is than I expected; of how healthy the plant communities seem; and of the sheer wildness of this system. These impressions don't lull me into forgetting that this landscape bears particular attention, given the impending energy development and other natural-resources extraction that will fragment this corridor for the large carnivores. And the intense patterns in this landscape—that is, the vegetation patterns tied to what the dominant herbivore is—leave me with more questions than answers.

While nature ultimately shapes trophic cascades, humans play a leading role in these relationships through the way we live on the land. I reflect on the importance of government regulations that control the predator-prey populations we've been observing on our continental journey. Environmental laws differ between the United States and Canada, and among states and provinces. In the next chapter, we'll see how these public policies can leave signatures on food web relationships and connectivity along the Carnivore Way.

CHAPTER THREE

Crossings

Pluie the wolf (*Canis lupus*) taught us that carnivores need room to roam. The details of how she lived from the time she was radio-collared in Kananaskis Country, Alberta, in 1991 until her death in 1995 illustrate the importance of carnivore conservation across political boundaries. For two years, biologists tracked her as she ranged over an area 40,000 square miles in size—ten times greater than Yellowstone. In the process, she moved through Banff National Park, where she was totally protected, to British Columbia, where she was liable to being hunted by humans. From there, she traveled south into the United States through Glacier National Park, where she was legally protected, and then into Montana, Idaho, and Washington, where environmental laws continued to protect her from being hunted or harmed. In 1993, she returned from the United States to British Columbia. There, a hunter's bullet nicked her collar, but didn't kill her—nevertheless, Pluie's collar stopped transmitting signals. Beyond that point, researchers didn't know whether the collar battery died or, indeed, what had become of her. Two years later in Invermere, British Columbia, a hunter legally killed her and her mate and their three pups.

As a researcher, I had a similar experience, although this hardly played out on such a large scale or as publicly as did Pluie's tale. In February 2007, for my trophic cascades research project, former Alberta Sustainable Resources Division biologist Carita Bergman and Waterton Lakes National Park ecologist Rob Watt radio-collared a wolf. A charcoal-gray yearling male, he was in good flesh and outstanding health. The data from his Argos satellite collar quickly showed his strong inclination to travel. The day after he was collared, he moved 80 miles south across the US-Canada border, into Glacier National Park, Montana, along with the Belly River pack. His activity in the southern half of the Crown of the Continent Ecosystem would provide a window into the challenges posed by transboundary conservation.

For a few days, the young wolf bounced around Glacier, spending time in the mountains, apparently hunting mountain goats (*Oreamnos americanus*) and bighorn sheep (*Ovis canadensis*). This was unusual behavior, because from central Alberta and British Columbia to central Arizona and New Mexico, elk (*Cervus elaphus*) are the wolf's preferred prey, followed by deer (*Odocoileus* spp.). The wolf next traveled down to the Blackfeet Reservation, moving along the Rocky Mountain Front and cutting through aspen (*Populus tremuloides*) foothills and rolling shortgrass prairie, where vast elk herds congregate in winter. While in Glacier, the wolf had "endangered species" status, which meant he couldn't be hunted or harmed. While on Blackfeet land, he also had "endangered" status, but this was based on tribal and federal wildlife laws, allowing actions such as wolf "removal" in response to livestock depredation, but no hunting or trapping back then.

Toward the end of the first week of being collared, the young wolf began to zigzag between the United States and Canada, spending his time mostly on ranchlands. While in Waterton, he had full protection from being hunted for sport, trapped for his fur, or shot as a predator. However, in Alberta beyond Waterton he had no legal protection. Wolves have never been protected in this province, perhaps because their population is robust overall, and they never have been extinct here.

Throughout the period of this wolf's journey, I was ensconced in my Oregon State University College of Forestry lab, working on my doctoral studies. "Please don't be a bonehead," I'd admonish the wolf every time I received data on my computer—a red dot vividly marking each location on a Google Earth map, the dots strung together in a way that indicated he was moving perilously close to ranchlands. I hoped he'd be able to steer clear of these potential trouble spots. However, on the tenth day after he was collared, an Alberta rancher shot him dead. The wolf hadn't been doing anything bad—hadn't eaten calves or harassed livestock, like some wolves do. But in Alberta a landowner or grazing leaseholder is legally entitled to shoot wolves within eight kilometers of his or her land, without a license.[1]

As a researcher, I was devastated by this wolf's death. However, I knew he had just been doing what wolves do—using a landscape exactly as others of his kind had done for thousands of years. Perhaps he was too young to understand the dangers of humans with guns. He certainly had no awareness of all the jurisdictions he had passed through, although more mature wolves sometimes begin to get it and avoid areas of human use. The policy implications of this young wolf's life and death demonstrate the power of environmental laws to shape animal lives. And his brief life made me wonder about the fate of other large carnivores that cross boundaries.

C8 80

In this chapter, I discuss international environmental laws that apply to large carnivores. To wolves crossing back and forth between Canada and the United States, or jaguars (*Panthera onca*) dispersing from northern Mexico into Arizona, borders have very real life-and-death implications, whatever the animals' awareness of them. For those of us, whether managers or concerned citizens, who are interested in conserving carnivores, it's important to be aware of these laws in the United States, Canada, and Mexico, and how they apply to dispersing animals. Environmental law is part of public policy—the system of laws that govern all our actions in the public sphere, which includes the natural world.

Like all laws, the statutes that dictate how we manage our environment can seem dry and technical. Nevertheless, they create a vital framework that touches all living things. You could say that environmental laws affect an animal like a lynx (*Lynx canadensis*), for example, just as tangibly as does the availability of its favorite food—the snowshoe hare (*Lepus americanus*).

As of late 2013, five out of the six carnivore species that I profile in this book are facing legal challenges to their protected status in the United States. For each of them, the federal government has filed a variety of legal documents in the *Federal Register* (a daily newspaper published by the National Archives). These documents furnish the basis for most public comments made regarding endangered-species management. To conserve animals such as the young male wolf whose progress I followed from Oregon State University, we need to understand what happens when an animal like him crosses legal jurisdictions. In that sense, environmental laws are living documents and powerful tools that can help conserve the web of life. While all these laws matter in the big picture of carnivore conservation, this chapter also can be used as a primer to learn about specific laws.

Policy expert Charles Chester defines *transboundary conservation* as international policies that focus on borders that present conservation challenges.[2] Within the context of carnivore conservation, issues along these borders have to do with the fact that nations share resources but don't always agree on how to manage the living species that they share. For example, a lynx dispersing from Colorado to Alberta will find itself subject to being legally trapped for its fur north of the US-Canada border. Canada bases its decision to "manage" this animal as a fur-bearing species and an economic resource (via the sale of its pelt) on the fact that many lynx live in the western portion of that nation, so there's no danger of this species becoming extinct there. Yet in the United States, that same lynx is staunchly protected by law because scientists have identified factors that could lead to its extinction.

But beyond environmental laws, today we're learning that it takes a village to conserve large carnivores, plus full awareness of the tools and challenges along the way. Nongovernmental organizations (NGOs), such as Yellowstone to Yukon (Y2Y), Wildlands Network, the International Sonoran Desert

Alliance, Naturalia, the Malpai Borderlands Group, and others, have become important agents of conservation across boundaries. However, this collaborative network is relatively new.

International Perspectives on Wildlife Conservation

Transboundary wolf stories graphically illustrate that large carnivores make big movements, crossing political boundaries and a variety of jurisdictions and land ownerships in the process. In the United States and Canada, this can include federal, state or provincial, tribal, and private land ownerships. Policies vary even within a particular jurisdiction. For example, the federal government manages national parks for preservation of nature, but manages national forests for natural-resources extraction and sustained yield (e.g., timber harvest and hunting). Adding further complexity, Mexico has communally owned lands, called *ejidos*. Established after the 1920 Mexican Revolution to take land from wealthy landowners and redistribute it to landless *campesinos*, or peasants, *ejidos* helped eliminate social inequalities.[3] Additionally, in the past decade, NGOs have become stakeholders in the United States, Canada, and Mexico, by purchasing and managing wildlife habitat, acquiring conservation easements, and working with government agencies to create cooperative wildlife management plans.

This mosaic of land ownership and policies means that carnivores traversing political boundaries within and between nations encounter different levels of protection. Table 3.1 provides a comparison of environmental legislation in the three nations as it applies to large carnivores in western North America. The following policy overview highlights leading environmental laws in the three nations and describes their strengths, weaknesses, and relevance to the large carnivores.

Two key international laws apply to large carnivore conservation. In 1963, the United Nations drafted the Convention on International Trade in Endangered Species of Wild Fauna and Flora (CITES), which covers international trade in endangered species. While it doesn't focus specifically on conserving endangered species, over the years, it's helped regulate their management among participating nations. The North American Free Trade Agreement (NAFTA) of

Table 3.1

Federal Legal Status of Six Large Carnivore Species in Western North America—
Canada, the United States, and Mexico

Species	Western US Status	Canadian Rocky Mountain Status	Mexican Sierra Madre Status
Grizzly Bear (*Ursus arctos*)	Threatened, relisted in 2009; delisting proposed in 2013	Special concern	Extinct, not protected
Gray Wolf (*Canis lupus*)	Threatened in Washington and Oregon; delisted in the Rockies; national delisting proposed in 2013, except for the Mexican gray wolf (*C. l. baileyi*)	Not protected	Endangered
Jaguar (*Panthera onca*)	Endangered since 1997; recovery plan pending in 2013	N/A	Endangered
Lynx (*Lynx canadensis*)	Threatened throughout the West since 2000; recovery plan pending in 2013	Not protected	N/A
Wolverine (*Gulo gulo luscus*)	Proposed Threatened status in the northern Rocky Mountains in 2013	Special concern	N/A
Cougar (*Puma concolor*)	Not protected federally	Not protected	Not protected

1994, which eliminated trade barriers between the United States, Canada, and Mexico, also regulates commerce in wildlife products. Both laws make it illegal, for example, to hunt a jaguar in Venezuela, where jaguar hunting is permitted, and then export its pelt to the United States as a big-game trophy. Until these laws were passed, many animals died in other nations in order to provide trophies for people living in the United States and Canada.

Since 2006, the International Union for Conservation of Nature (IUCN) has complemented CITES. The IUCN Red List identifies species most in need of international conservation and reports changes in their status. Red List species status ranges from "extinct" to "threatened" to "at risk." However, this list doesn't provide actual protection to species; rather, it functions as a guideline for their international management.[4]

US Environmental Laws

A spate of powerful environmental legislation emerged in the 1960s, inspired by the writings of Aldo Leopold and Rachel Carson. In a sense, these laws can be seen as the precepts of Leopold's land ethic: "A thing is right when it tends to preserve the integrity, stability, and beauty of the biotic community," as well as Carson's outrage about how we were fouling nature with toxic chemicals. The two resulting laws with the greatest bearing on large carnivore conservation in the United States are the National Environmental Policy Act (NEPA) and the Endangered Species Act (ESA). In 1969, NEPA mandated that, prior to taking actions that may significantly affect the quality of the environment, public and private land managers must produce a report called an Environmental Impact Statement (EIS). This planning-process law, which governs human land uses on federal lands, has become a key tool in preventing extinction.[5]

The 1973 ESA protects species that risk extinction. To do so, this law gives imperiled species "endangered" or "threatened" status.[6] While Congress (the legislative branch of the US federal government) makes environmental laws, the US Fish and Wildlife Service (USFWS), which is part of the executive branch of the federal government, administrates the ESA via the rule-making process. USFWS files rules in the *Federal Register* that represent the agency's interpretation of the ESA. Putting a species on the USFWS list of threatened and endangered plants and animals is part of this rule-making process, legally and commonly referred to as "listing" a species. The ESA defines an *endangered species* as one "in danger of extinction throughout all or a significant portion of its range," and a *threatened species* as one "likely to become an endangered species within the foreseeable future." Additionally, it addresses conservation of the habitat that species need to persist and thrive, termed *critical habitat*. Notable ESA successes include the bald eagle (*Haliaeetus leucocephalus*) and peregrine falcon (*Falco peregrinus*).[7] Like all laws, the ESA has strengths and weaknesses.

Its strength lies in its flexibility, which includes provisions that acknowledge landowner rights, provide incentives to voluntarily conserve listed species, and improve wildlife stewardship (e.g., Safe Harbor Agreements and Candidate Conservation Agreements). The Section 10(j) Nonessential Experimental Population Amendment gives agencies flexibility to address public concerns and to minimize conflicts created by reintroduced species, such as wolves, by allowing controlled lethal removal under certain circumstances.[8]

The ESA's weaknesses include its cumbersome application and vague terminology. It takes many years for species to be listed. In the meantime, those being considered for listing, called *candidate species*, receive no protection.[9] The wolverine (*Gulo gulo luscus*), a species with very low numbers and rapidly shrinking habitat due to climate change, has taken years to list. The ESA's fuzzy language regarding designation of a species' population for protection, termed a *distinct population segment* (DPS), is also problematic.[10] Lack of clarity about this term's meaning and application has led to multiple lawsuits over the years. Additionally, the ESA provides inadequate funding for monitoring.

Another weakness has to do with defining *recovery of a species*. While the ESA does not offer a definition of what this means, ecologists generally think of recovery as persistence of a species with sound genetic diversity for at least 100 years. The actual number of individuals is not a one-size-fits-all prescription, and is based on individual species' natural history traits (e.g., the number of young it produces). ESA recovery plans, which are filed as "rules" in the *Federal Register,* contain recovery criteria that are tailored to individual species. For example, the recovery criteria for long-lived species with a lower reproductive rate may be more stringent than those for shorter-lived species that have higher reproductive rates. As we shall see in the species chapters that follow, our ambiguous definition of *recovery*, which leaves much room for interpretation, can make it difficult for species to thrive. Often, recovery thresholds are based more on "social carrying capacity," or what human communities will tolerate, than on the biological needs of a species in order for it to achieve scientifically defined recovery.

The ESA's critics have requested revision to make listing species more difficult and to eliminate critical-habitat designation. Currently, the ESA calls for listing based on a species being in danger of extinction throughout all or a "significant portion" of its range. A 2012 proposal by the USFWS calls for redefining the meaning of "significant portion."[11] This could allow more liberal interpretation of the ESA and, consequently, could reduce protection for species.

All laws fall into two categories: procedural and substantive. *Procedural laws* define actions that can be taken and the procedures required to take those actions. NEPA is a procedural law that calls for developing an EIS when actions to be taken (e.g., timber harvest) might affect a threatened or endangered species, such as the lynx. *Substantive laws* take things further: they prohibit specific actions, in effect functioning like legal "big dogs." Few substantive environmental laws exist globally. The ESA, a very big dog with very sharp teeth, is filled with substantive prohibitions. Jointly, the ESA and NEPA have tremendous power to abate or stop actions such as habitat destruction (e.g., timber harvest in lynx territory), which can drive species to extinction.[12]

The ESA's roots in science make it an even more remarkable law. According to environmental lawyer Doug Honnold,

Again and again in the ESA, there's the provision to use the best available science. Consequently, the ESA is ever new. When new science tells us what we need to build by way of a roadmap to protect species, the ESA says don't wait, put it right into play. If you look at environmental statutes, few have that kind of language today, and precious few had that kind of language back in 1973.[13]

However, the world has changed since 1973. According to Honnold, back then,

. . . we had this innocent notion of, "Ah, Noah's Ark. We can get everybody on the boat and there'll be no problem." Now we've got a species-extinction crisis that, even with all available resources, would be difficult to address. This has created the need for triage. Climate change makes all systems exponentially unpredictable, which then makes it very difficult to ensure that long-term planning will be effective. This means that, for a significant number of species, there's going to be a need for some level of human intervention, such as assisted migration. We're living in a brave new world that we hadn't foreseen in 1973.[14]

In the United States, private citizens and NGOs can sue the government if they disagree with the way environmental laws are being interpreted. As a result, over the years, lawsuits have improved implementation of these laws, refined our understanding of them, and, in many cases, led to increased protection for species. For example, when USFWS failed to protect the jaguar in the United States due to a legal technicality, conservation organizations filed lawsuits. Consequently, today the jaguar has ESA protection.

When a species is listed under the ESA, the next step calls for creating a recovery plan. This science-based plan must be peer reviewed and approved by Congress. Such plans define the criteria for recovery and include provisions for *downlisting*, which reduces a species' status from "endangered" to "threatened," and *delisting*, which removes most protection. Recovery plans address issues such as habitat, genetic diversity, and human interests. An important step in this process calls for federally approved state management plans. Upon delisting, the states take over management of a "recovered" species. Depending on the species, such plans can include hunting. However, USFWS continues, post-delisting, to monitor species.[15]

To use the gray wolf as an example, in 1974, Congress listed this species as "endangered" under the ESA. Next, the federal government established the Northern Rocky Mountain Recovery Area (NRM), which included Idaho, Montana, and Wyoming, and began to create a recovery plan for this region. The plan, which took twelve years to create, called for wolf reintroduction in Yellowstone National Park and central Idaho. To provide flexibility in dealing with problem wolves, the plan gave this population "nonessential experimental" status, via Section 10(j) of the ESA.[16] The 1995–1996 NRM reintroduction of 66 wolves was an unprecedented success. By 2002, this population had met recovery goals of at least 300 wolves and 30 breeding pairs in the three NRM states for at least three consecutive years, so USFWS downlisted wolves from "endangered" to "threatened."

Prior to federally delisting wolves, Idaho, Montana, and Wyoming had to create management plans to ensure that the species wouldn't go extinct. Idaho and Montana produced sound plans. The Wyoming state legislature produced a plan, deemed unacceptable by USFWS, that classified the gray wolf as a trophy game animal in wilderness areas and as a predator not subject to management or protection elsewhere. The plan failed to pass federal muster because its harsh measures (e.g., broad-scale wolf killing everywhere but in wilderness areas) could drive the wolf back to extinction in that state. Over the next five years, Wyoming worked to improve its plan.

USFWS issued the first NRM delisting proposal in 2007. It defined a distinct population segment (DPS) that included Idaho, Montana, and Wyoming, plus eastern Washington and Oregon. A federal court found this DPS unacceptable. In April 2009, USFWS attempted to delist the NRM DPS, without Wyoming. The court again found this unacceptable, because the ESA doesn't permit delisting a DPS incrementally. However, Congress delisted NRM wolves in 2011 via a Department of the Interior appropriations bill, which reinstated the 2009 delisting. Wolves have been delisted in the NRM since then, with hunting allowed in the three states.[17]

In the United States, individual states regard themselves as trustees of wildlife. Most have their own endangered species acts, supplemental to the ESA. However, states don't have the right to reduce or ignore ESA protection. Management objectives among states vary, reflecting cultural differences, but are based on extractive uses such as hunting and trapping. In the northern Rocky Mountains, for example, Montana has a state endangered species act, but Idaho and Wyoming do not. Each state has its own management regulations for non–federally protected wildlife. States define "predators" differently and, post-delisting, have different policies about wolf control, as we've seen with Wyoming.[18]

Canadian Environmental Laws

Canadian environmental legislation, while not as powerful as that in the United States, is fairly well developed. The Canadian Environmental Protection Act (CEPA) of 1992 is the main Canadian environmental law. It regulates toxic substances, environmental pollution, and environmental management. The Canadian Endangered Species Protection Act, introduced in 1996 and widely criticized for being a flawed statute, "died on the Order Paper" (to use Canadian legal parlance). This means the bill didn't reach the final stages of becoming a law, so Parliament terminated it. In 2002, the federal government passed the Species at Risk Act (SARA), a revised draft of the earlier law. SARA aims to prevent extinction and provide for species recovery and management. The Committee on the Status of Endangered Wildlife in Canada (COSEWIC), a group of scientists, evaluates species for listing under SARA.[19]

Like the ESA, SARA focuses on how human activity affects species conservation. Also like the ESA, SARA designates imperiled species as "endangered" or "threatened," but adds a third category, "species of special concern" (defined as one that may become a threatened or endangered species in the future). However, SARA lumps all of these categories under the "species at risk" umbrella category. It prohibits actions that kill, injure, or interfere with endangered species, and it requires recovery plans.[20]

SARA's critics claim that it contains arcane mechanisms for enforcement and falls short of what's needed to prevent extinction and to ensure full recovery of species. Corporations and federal entities can be held accountable for SARA violations. However, SARA's leading weakness (and the way it differs most from the ESA) has to do with the fact that, in Canada, citizens and NGOs have very limited ability to sue the federal government to improve how environmental laws are being implemented.[21]

As is the case in US states, Canadian provinces also have their own environmental legislation. Two Canadian provinces lie within the Carnivore Way—Alberta and British Columbia. In Alberta, the Wildlife Act of 1987 and the Wildlife Amendment Act of 1996 allow provincial governments to establish wildlife conservation regulations, including identifying endangered species. British Columbia doesn't have endangered species legislation. However, its Ecological Reserve Act (1979), Park Act (1979), and Wildlife Act (1982) can protect sensitive species.

The grizzly bear (*Ursus arctos*) provides a good example of how Canadian and US environmental laws can be applied across provincial and international boundaries. In Canada, SARA has given this species "at risk" status.

However, the strongest protection can be provincial. In British Columbia, a grizzly bear can be hunted, while in Alberta, this species has provincial "threatened" status, so hunting it isn't allowed. Like other carnivores, the grizzly covers a lot of ground. Let's say that a four-year-old male grizzly is born in northwest Montana. During the first few years of his life, he has total protection and can ramble about freely without worrying about meeting a bullet, as long as he doesn't get into people's food or kill cattle. But at the beginning of his fourth year, he awakens from hibernation feeling restive and leaves to get away from some of the adult male bears who've begun to pick on him. He travels through the North Fork of the Flathead Valley, toward Canada. When he reaches the US-Canada border, he has several options. If he goes north or west, he'll find himself in British Columbia, in more big mountains. If he goes east, he'll find himself in Alberta, in mountains that quickly turn to grasslands. Although he doesn't have any awareness of the legal niceties involved in his choice, his travel route could mean the difference between life and death; in Alberta grizzly bears are strictly protected, while in British Columbia, they can legally be hunted.

Mexican Environmental Laws

Of the three nations along the Carnivore Way, Mexico has the weakest environmental laws, due, most likely, to economic stresses. Mexico has long participated in international conservation. However, although a signatory to CITES and NAFTA, illegal trade in wildlife and wildlife products continues in this nation (e.g., smuggling parrots to the United States).

Recent Mexican environmental legislation includes the 1988 General Law of Ecological Balance and Environmental Protection (also known as the Federal Ecology Law). This law attempts to balance human needs with nature conservation, and it includes environmental impact assessments. The official Mexican environmental norms (*normas oficiales Mexicanas*, or NOMs), aid in implementing this law. Additionally, the 2000 General Wildlife Law (Ley General de Vida Silvestre, or LGVS) sets standards for sustainable use of wildlife, land stewardship, and habitat restoration. Most notably, the LGVS identifies species at risk. However, limited mechanisms and funds exist for enforcing these laws, and they can't be litigated by citizens or NGOs.[22]

Tribes and Environmental Law

None of the indigenous/aboriginal people in the United States, Canada, and Mexico are independent of federal laws. In the United States and Canada, indigenous status conveys sovereignty, via treaty rights. Honoring these rights requires coordination among federal and state governments and the tribes or First Nations involved.

As European settlement progressed in the United States in the 1880s, tribes were pushed onto reservations. Via treaties, these tribes retained or reserved rights that they had always possessed, including some rights that non-Indians don't have. For example, Indians have the right to engage in subsistence hunting, which comes with far more liberal regulations than does hunting by non-indigenous people. Indians have control over activities that take place within their reservation, especially activities that maintain their culture. However, this doesn't exempt them from state or federal laws. States have the right to protect state-listed endangered species on tribal land and to control activities that may jeopardize these species.[23]

In Canada, First Nations similarly retain aboriginal rights via treaties. As in the United States, federal environmental laws supersede such aboriginal rights. Environmental laws supersede most other laws, even tribal laws. While the scope of sovereign rights remains to be settled in many parts of Canada, treaty rights and land-claim agreements give aboriginal people the right to participate in land-use planning and environmental assessment. Additionally, the federal government must implement environmental laws in a manner that honors traditional aboriginal interests. *Westar Timber Ltd. v. Gitskan Wet'suwet'en Tribal Council* provides a good example of this. A British Columbia aboriginal community protested road construction that would cut down an old-growth forest on their land in order to facilitate access to timber. The logging company argued that not building the road would cause the loss of jobs and revenue.

The tribe argued that the pristine forest to be eliminated by the road had high cultural value for them. The court awarded the tribe the injunction that they sought, and the road wasn't built.[24]

Mexico gives no formal legal recognition to indigenous people. Rather than a reservation system, indigenous people in Mexico retain *ejidos* (communal land that community members possess and use for agriculture). However, while indigenous people have the right to use these lands, they don't have any formal legal jurisdiction over them. As in the United States and Canada, indigenous Mexican communities make decisions communally, via tribal councils. While the federal government doesn't uniformly recognize their rights, the Federal Ecology Law represents the best tool for implementation of them.[25]

International Challenges to Large Carnivore Conservation

The national and international laws described in this chapter provide an imperfect safety net: their limited vision creates challenges for large carnivore conservation. For example, based on now-dated science, the laws fail to acknowledge and apply what scientists have known since the 1980s about the ecological role of keystone species. The ESA and SARA offer protection for species based on rarity. However, as conservation biologist Reed Noss points out, "rarity is a poor indicator of ecological role or importance."[26] Wolves can indirectly improve habitat for a variety of other species, such as endangered native trout, via the trophic cascades mechanism of enabling stream vegetation to grow. This vegetation, in turn, shades streams and lowers water temperature. Such scientific knowledge suggests a need to revisit how we apply the ESA.

Philosophically, the three nations have compatible environmental laws. However, the key factor limiting a more integrated approach across boundaries has to do with the very different mechanisms available in each nation for implementing and enforcing those laws. CITES provides a unifying mechanism, but by itself this is insufficient to ensure persistence of a species such as the jaguar across boundaries. Consequently, NGOs such as Naturalia have been taking a proactive role by taking actions such as creating jaguar reserves on private lands near borders.

In all three nations, laws that protect wildlife give precedence to economic development. In many cases, the national government has accommodated special interests, such as energy, timber, mining, and ranching industries, jeopardizing species at risk. For example, in the case of the grizzly bear, in British Columbia ample science shows that this species may be more at risk of extinction than is generally reported by provincial managers. However, outdoor outfitters insist on continuing to hunt the grizzly because of the revenue that trophy hunts bring in. Consequently, since 2004, there's been a heated debate in that province about whether managers are applying the best science.

The 1,900-mile US-Mexico border graphically illustrates the challenges of international wildlife conservation. In 2006, President George W. Bush authorized a fence along this border to reduce illegal drug trade and illegal immigration. As of 2010, the federal government had built 700 miles of border fence from San Diego, California, to Texas, just east of El Paso, positioned at crime hotspots. The discontinuous fence consists of a variety of barriers, including vehicle barriers permeable to humans and wildlife. Figure 3.1 depicts the most severe barriers, which consist of eighteen-foot and thirteen-foot steel beam fences, designed to block human pedestrian traffic—which also block wildlife passage. John Davis's 5,000-mile TrekWest for Wildlands Network from northern Mexico to Canada took him across the US-Mexico border in an area that was fenced. He refers to the border fence as "an abomination that has created the worst threat to connectivity for wildlife in North America."

The US federal government has acknowledged the problems created by the fence. In 2009, the US Department of Homeland Security committed 50 million dollars to mitigate environmental damage done by the fence. And in 2010, President Obama halted fence construction due to the unsustainable, multi-billion-dollar expense of completing and maintaining it. As of 2013, the fence hadn't effectively stopped illegal human activities, but it had effectively blocked passage of endangered species, such as jaguars.[27] Barriers such as this thwart the IUCN's stated global priority of conserving genetic diversity. Animal populations that have high genetic diversity are more resilient to changes in their environment. Regardless of the species, small populations cut off from others can go extinct.[28]

Figure 3.1. Steel beam fence blocking deer from crossing the US-Mexico border. (Photo by Defenders of Wildlife.)

Jaguar and cougar (*Puma concolor*) conservation along the US-Mexico border provides a good example of how transboundary conservation might be improved. Here, I focus on the border segment separating the states of Sonora and Chihuahua in Mexico from Arizona and New Mexico in the United States. Scientists have identified this as an important area for connectivity. It contains most of the currently constructed border fence, so it has permeability issues. Ranching represents the dominant land use here.[29]

Northern Mexico has a low jaguar population, while Arizona and New Mexico have no established populations of this species, just occasional dispersing males. In Mexico, jaguars have federal protection, but they can be moved by managers when conflicts arise with ranchers. The federal government sometimes issues permits for legal hunting of jaguars in order to remove animals that prey on livestock. Nevertheless, ranchers kill jaguars illegally. In the United States, the federal government strictly protects jaguars. Maintaining a connected jaguar population across the international border is critically important to the long-term (e.g., more than 100-year) survival of this species in both nations.

Cougar conservation status is similar along both sides of the US-Mexico border. In Sonora, Chihuahua, Arizona, and New Mexico, hunting policies, albeit liberal, maintain a relatively high cougar population. However, on both sides of the border, but especially in Mexico, ranchers legally kill cougars to protect livestock. Maintaining a connected cougar population across the international border is important, but not crucial, to the survival of a species present in such high numbers.[30]

Jaguar and cougar conservation involves a joint effort. As we've seen with Y2Y along the US–Canada border, efforts must begin at local scales within critical habitats, but they must also take place at state and federal government levels and across international boundaries. Support from the US Department of Homeland Security, for example, can help advance jaguar conservation. Existing maps of jaguar and cougar movements can inform connectivity planning. Many organizations and institutions in Mexico and the United States can support transboundary collaboration, but these resources have been underutilized. Nevertheless, informal collaboration between NGOs and stakeholders is beginning to lead to more-formal international policy. For example, in early 2012, USFWS assembled a binational Jaguar Recovery Team composed of Mexican and US experts. This team, including ranchers from the United States and Canada, is advising the governments of both nations on how to recover jaguars and improve jaguar–human coexistence.[31]

El Norte and the Geography of Hope

I am intimately familiar with transboundary issues. My parents were born in northern Mexico, near the US border. They came from ranching backgrounds. My maternal great-grandfather and his brothers owned a ranch that encompassed a good portion of what is today the city of Juarez, Chihuahua. My paternal grandfather owned a several-hundred-thousand-acre ranch in Chihuahua near the Sonora border, in the Sierra Madre Occidental—the same area Aldo Leopold visited in the 1930s on a bow-hunting trip and described in the original foreword to *A Sand County Almanac*. About this landscape he wrote, "It was here that I first clearly realized that land is an organism, that all of my life I had seen only sick land, whereas here was a biota still in perfect aboriginal health."[32] In Mexico, the far north is still regarded as the frontier. Well into the twentieth century, *La Frontera*, or *El Norte*, as this outback is called, harbored fugitive Apaches and relict jaguars and wolves.

When he wasn't away at boarding school in the 1930s, my father worked on his family's ranch. His job was to guard the cattle and kill all the predators he saw—especially wolves. Toward the end of his life, he confessed to me that he could never bring himself to shoot predators, particularly wolves, because they never harmed the cattle.

The world has changed much since my father was a young man. My family's Juarez ranch has since been developed as real estate, carved up into increasingly smaller parcels. My family lost their ranch in Chihuahua for financial reasons. The story goes that my grandfather won the ranch in a poker game in the 1920s and lost it 25 years later . . . in another poker game. Heartbroken about the loss of the ranch, my father went on to a successful career in the Mexican diplomatic service. He specialized in international trade and social justice. When I was a child, our dinner-table conversations were often about trade, international law, natural-resources extraction and how that affected people, and the thorny immigration problems at borders.

I grew up in other nations, including Canada and the United States, but I always returned to Mexico for part of each year. My transboundary education heightened my awareness of what it takes to conserve whole landscapes. It also gave me a feet-on-the-ground understanding of how cultural norms and economics, which differ among nations, lie at the core of transboundary issues.

Today, my family's former Chihuahan ranchlands have been ravaged by drug cartels. Nevertheless, those lands defy dewilding. The Mexican government has implemented fledgling environmental laws and policies to protect large carnivores. Wolves and jaguars, their numbers quietly growing, currently live in the *barrancas*—the remote rough canyons—and the *arroyos* in this landscape. When TrekWest took John Davis through the heart of El Norte, he found these lands surprisingly intact.

What I experienced at my California home or my family's Juarez ranch is happening everywhere. Between unbridled real estate development, a burgeoning human population, and aggressive gas extraction, from Alaska to Mexico we're plundering the pelf of the West. Even so, the Carnivore Way contains the last redoubts of wildness, places where one can still fall asleep to the sound of wild wolves howling and find jaguar tracks pressed into streambank mud.

American writer Wallace Stegner alerted us to how much we need wild country and carnivores. He said that such wildness "can be a means of reassuring ourselves of our sanity as creatures, as part of the geography of hope."[33] Environmental laws in the United States, Canada, and Mexico are a big part of why we still have large carnivores roaming remote mountain valleys.

During his final years, my father and I read Leopold and Stegner together. Not long before he died, he told me that he never fully understood his affinity for large carnivores. But back in the 1920s and 30s, he instinctively knew what Leopold, Stegner, and others also knew —long before we had trophic cascades science—that saving wildness across boundaries was ultimately about saving ourselves. My father died with hope that we would.

In the carnivore chapters that follow we'll explore the intimate connection these big, fierce creatures have with the landscapes in which they live. And we'll see how both the carnivores and the landscapes they roam are part of the geography of hope.

PART TWO

Where the Carnivores Roam

Grizzly Bear (*Ursus arctos*). (Drawing by Lael Gray.)

CHAPTER FOUR

Grizzly Bear (*Ursus arctos*)

If you go to a place often enough, eventually it claims you. If you sit there long enough and still enough, sometimes the creatures who live there stop being able to tell the difference between you and one tall tree. That's when things get interesting.

My writing cabin, a homestead, is an inholding in a northwest Montana national forest. Several trout streams run through our land. I like to sit on the bank of one of them, under a lone larch tree, and write. In doing so, over the years, I've gotten to know the sounds of the resident wildlife: the sweet lilt of the yellow warbler and the buzz-song of the varied thrush; the *basso-profundo* breathing of moose browsing in the willows on the far side of the creek. These sounds come together layer on layer, overlain by the soughing of the wind through the lodgepole pines.

I heard all of these sounds one sultry June afternoon as I sat writing poetry. I didn't look up, chiefly because I didn't want to risk making eye contact with a moose with a calf at her side. As I sat there quietly, a light wind blowing toward me, the forest animals probably had no idea I was there.

All at once, I heard something different. Even as I was lost in my poem, this sound's dissonance awakened the ancient part of my brain that deals with instinct and survival. There was a deep, heavy rustling, something big and soft-footed pawing through the cattails that grow in front of the willows and right up to the creek, which I should add, is a scant ten feet wide.

I looked up and made eye contact with a grizzly bear, about twelve feet from where I sat, my back pressed against the larch's furrowed bark. A large juvenile, probably a three-year-old, with rounded ears, brown eyes, and coppery, silver-tipped fur. Thirsty on this hot day, the bear had come to the creek for water. We looked at each other for several heartbeats, no fear on either side. Beyond words, beyond time, we continued to take each other's measure. Then both of us had the same thought at the same time. In slow motion, we got up and calmly turned and walked away in opposite directions. An amicable parting.

A hundred feet from where I'd seen the grizzly, my twenty-first-century brain re-engaged, and I realized what had just happened. There I'd been, without the protection of pepper spray (a deterrent that can stop a charging bear in its tracks), face-to-face with a two-hundred-pound grizzly. The strange thing is, I didn't feel frightened, even then. Moving slowly, I went to get pepper spray. My husband Steve, who was doing some construction on our cabin, asked, "Anything interesting at the creek?" I mumbled "Uh-huh," grabbed the spray from a shelf next to the door, and returned to the creek, a woman on a mission to reclaim her favorite writing spot.

When I reached the creek, the young bear was nowhere to be seen. I sank down under the larch and finished my poem. After that, I returned to our cabin and told Steve what had happened. I expected a lecture about foolish risks. Instead he just smiled and said, "Cool." And again I was reminded of why I married this man.

In the years since, I've made many visits to that larch and haven't seen that bear again. Nevertheless, he left me with a deeper understanding of what it means to live peacefully and respectfully with bears.

CʒFꙎ

Our tangled history with the great bear ranges from fearing and persecuting it to renewing the bonds we had with it before the European settlement of North America. As I delved into grizzly bear literature to write this chapter, the theme of bear–human conflicts kept coming up. To better understand these relationships, I visited Alaska, where the human population drops and the great bear and other predators abound. My niece and nephew Alisa and Andrew Acosta and environmental writer Donna Fleury accompanied me on this portion of my Carnivore Way travels.

Grizzly Bears in Big Wild Places

In Alaska, we stayed with Tom Meier, who led Denali National Park's biological program. He and I first met in the early 2000s, when I tracked wolves (*Canis lupus*) for him in Montana in order to document a dispersing wolf population. At the time, he worked for the US Fish and Wildlife Service (USFWS, or just "the Service") on the Montana wolf reintroduction, and he encouraged me to attend graduate school.

Denali National Park and Preserve encompasses 6 million acres. Congress established it in 1917 at the recommendation of naturalist and hunter Charles Sheldon, who had spent time in the area between 1906 and 1908. Originally called Mount McKinley National Park, in 1980 Congress renamed it Denali ("The High One" in Athabaskan), expanded the park to its current size, and created an adjoining national preserve. National preserves are lands associated with national parks in which Congress allows public hunting and trapping. The US National Park System has eighteen national preserves, ten of them in Alaska.[1] A single road runs through about half of Denali's interior. Since 1972, this 92-mile road has been closed to private-vehicle traffic beyond the first fifteen miles from the entrance. To access more of the park, visitors can travel by bus, foot, bike, skis, or mush dogs.

In Denali, Tom has arranged for us to join him and Bridget Borg, a park wildlife biologist, to look for carnivore activity, with a focus on wolves. (This particular day, it so happens, will bring us insights on grizzlies as well as wolves.) Tom has reserved a park vehicle to drive us past the Savage River Bridge, where the restriction on private vehicles begins. Our first stop will be the East Fork of the Toklat River, where we'll visit some of Adolph Murie's study sites. A biologist from Minnesota, Murie conducted wildlife surveys in Yellowstone National Park in the 1930s. He came to McKinley Park in April 1939 to do scientific field research on wolf–Dall's sheep (*Ovis dalli*) interactions and ecology, and stayed through October. He returned in April 1940, and for the next fifteen months focused not only on wolves and Dall's sheep but also on other interrelated species and on the greater ecosystem, laying the foundation for today's large carnivore research in Denali and North America.[2]

Since its establishment as Mount McKinley National Park, Denali has fostered awareness of the significance of large carnivores to ecosystem health and conservation. In the 1930s, most people considered this a crazy notion. However, since Murie's time, public attitudes toward predators have been shifting from fear to respect. Drawn to wildness, many people today come to Denali to watch wolves and grizzly bears fulfill their ecological roles in an "intact ecosystem" (e.g., one with minimal human impacts and inhabited by all the species present before European settlement of North America). Carnivore interactions with their prey dominate Denali's food web. These far-reaching relationships entail not just wolves and their prey, but the other carnivores as well. Bears, for example, benefit directly from wolf predation on ungulates, which provides them leftover meat to scavenge. Murie was among the first to note these secondary relationships.

Clouds shift across the mountaintops as we reach the Savage River. A misty summer rain has been falling all morning. We turn southeast off the Park Road onto a narrow spur road that takes us to the East Fork Cabin—one of the cabins in which Murie and his family summered while he worked here. We park, ford a rushing creek, and then bushwhack through the shrubs along the braided East Fork River. The willows (*Salix* spp.) contain lots of bear sign: enormous tracks and piles of scat. Upon crossing the creek, I experience walking on tundra for the first time. A deep layer of sphagnum moss covers the ground. With each step, the moss gives beneath my feet, as if I were walking on a soft mattress.

As we make our way through willow and alder (*Alnus crispa*) thickets, we talk about bear encounters. I'm astonished that, between them, Tom and Bridget have had only one close encounter with a grizzly bear. This differs from

Alberta and Montana, where I bump into them often. I think about coexistence and how in a landscape as vast as Alaska, perhaps these relationships are simpler than down south. In the far north, which has a much lower density of both human and bear populations, the potential for conflict between bears and humans may be correspondingly lower.

We don't find fresh wolf sign, even though, historically, wolves have used this area extensively. Over lunch we talk about wolves (I discuss Tom's insights on wolves in Chapter 5), and then we return to our vehicle and explore farther along the Park Road. Continuing southwest through the Toklat drainage toward Eielson, we come upon a female grizzly with two cubs, eating forbs on the tundra amid a profusion of wildflowers. We stop to watch from a respectful distance of 300 yards. A couple miles farther, we spot a large male grizzly digging up the tundra. The grizzly bears here are a pale caramel color and more robust than their southern relatives. Even the cubs look larger here. They also have longer fur, likely due to the extreme cold. Later in the day, we see the same bear mother and cubs again, this time napping in a willow copse. She reclines on a moss bed, surrounded by white avens (*Dryas octopetala*) blossoms, creating one of the loveliest tableaus I've witnessed in nature. After photographing her, we slowly make our way out of the park, our senses filled with the gift of this day.

We have no way of knowing that four weeks from today Tom will be dead of a possible heart attack. Or that in five weeks the park's first fatal bear attack on a human will occur in the West Branch of the upper Toklat River, a few miles from where we just traveled with Tom and Bridget. The West Branch topography and plants are similar to those in the East Fork: wide gravel bars, a braided river, and lush willows and shrubs that provide ample forage for grizzly bears. Later, I'll struggle to assimilate the news of these two independent losses: my mentor's death and this tragic mauling.

What's in a Name: The Northern Horrible Bear

Taxonomy is the science of naming and organizing living things. As such, it provides the first clues about a species' history, nature, and needs. Taxonomists group organisms (e.g., carnivores) by shared traits, called *taxa*, which they then rank in a hierarchy. In 1758, Swedish scientist Carl Linnaeus developed the binomial nomenclature we still use today, giving each organism a genus and species name. Taxonomists use the *subspecies* to identify an interbreeding, geographically isolated population of a species.

Many colorful stories exist about how species acquired their taxonomical names. The grizzly bear, also called the brown bear, provides a perfect example. In the late 1700s, Linnaeus classified it as *Ursus arctos*, meaning "northern bear." But in 1814, based on accounts from the Lewis and Clark expedition of this bear's appearance and behavior, taxonomist George Ord renamed it *U. horribilis*. Some believe that Ord misunderstood the word "grizzled," which Lewis and Clark used to refer to the gray hairs in the bear's fur, and instead translated the word "grizzly" into Latin as "*horribilis*." However, historian Paul Schullery suggests that Ord, taken by Lewis and Clark's descriptions of the bear's fierce nature, deliberately chose the name "*horribilis*" as a pun on "grizzly"/"grisly." For the next century and a half, this unfortunate misinterpretation set the tone for European relations with this bear. Eventually, taxonomists renamed it *U. arctos horribilis*, which means "northern horrible bear."[3] As a professional group, taxonomists have their own subspecies: the lumpers and the splitters. The twentieth century was the heyday of the splitters, who used things such as skull shape to subdivide known species into finer and finer categories. At one time, taxonomists classified brown bears into 86 subspecies, largely the doing of Clinton Hart Merriam, the leading splitter of the last century. Today, we know that many erstwhile subspecies are genetically the same, and the lumpers are prevailing. While taxonomists seldom agree, they do recognize that the brown bear (*U. arctos*), is a *holarctic* species (i.e., one distributed around the world in its northerly latitudes) that is found across much of northern Eurasia and northwestern North America.

Grizzly bears arrived in North America via the Bering Land Bridge toward the end of the last ice age. They spread south 12,000 years ago as the continental ice sheets retreated. Currently, taxonomists acknowledge just two North American subspecies: *U. a. horribilis*, the inland grizzly bear found in interior Alaska, the Yukon, and the Rocky Mountains, and *U. a. middendorffi*, the larger Alaskan brown bear found along the coast.[4]

To confuse things further, we use common names for organisms. Common names can change from place to place, as they reflect local cultures. As is the convention in mammalogy, I use the name "grizzly bear" to refer to all the brown bears in North America generally, and "brown bear" for the brown bears that range specifically along coastal Alaska and British Columbia. In this context, the word "brown" doesn't refer to this species' color, but to this subspecies' genetics.

Grizzly Bear Natural History

Conservation of any species begins with awareness of its habits and needs. *Natural history,* defined as the study of animals and plants using observation to note life history traits, is one of the oldest forms of science. Two thousand years ago, first-century naturalist Pliny the Elder compiled the first account of wild animal traits in his *Historia Naturalis.* Naturalists have been building on this knowledge bank ever since.[5]

Grizzly bears are formidable-looking creatures. Varying in color from pale blond to nearly black, many have brown, white-tipped (grizzled) fur. In the interior of North America, adult female grizzly bears typically range from 250 to 350 pounds. Males are bigger, weighing between 400 and 600 pounds. Mammalogists call this size difference *sexual dimorphism.* However, in coastal Alaska and British Columbia, male bears can weigh up to 1,500 pounds. For the grizzly bear, *field marks* (a species' observable physical traits) include small, round ears, a dished-in face, round head, and pronounced shoulder hump. But their long, often pale-colored claws are probably their most intimidating field mark. They use their claws primarily for digging, as well as for catching and eating salmon. Indeed, in a River of Words Project poem, a third-grader described bears as having "claws like forks." In comparison, black bears (*U. americanus*), the species with which the grizzly is most often confused, have much shorter claws, longer snouts, and no hump. Both come in similar colors, so color isn't a good way to tell them apart.[6]

With one of the lowest reproductive rates of all land mammals, grizzly bears face significant conservation challenges. They live 25 to 30 years.

Females reach sexual maturity at between five and six years in the south, a bit later in the far north. They breed in late spring to early summer. However, it takes months for a fertilized egg to attach to the female's uterus, called *delayed implantation* or *embryonic diapause*. Grizzly bears developed this fascinating evolutionary adaptation to increase survival of breeding females. During poor food years (e.g., when shrubs don't produce enough berries), rather than both the female and her fetus dying, delayed implantation increases the female's chances of surviving, though at the expense of her fertilized egg. Females produce an average of two young every two to four years and are intensely protective of them. Adult males can prey on all cubs, including their own; females sometimes prey on other females' cubs. Therefore, females with cubs keep their distance from other adults. Cubs stay with their mothers for three years. Juvenile males disperse by leaving their mother's home range as subadults. Juvenile females tend to be *philopatric*, which means they remain within their mother's home range during adulthood. A male's home range can be from 400 to 1,000 square miles, and a female's from 80 to 200 square miles; however, home ranges can vary considerably, depending on food availability. Given their slow reproductive rate and female philopatry, grizzly bears aren't very ecologically resilient.[7]

While they belong to the order *Carnivora*, grizzly bears are in fact omnivores that eat anything from plants to nuts to other animals. In Yellowstone, they've been observed eating over 200 different species of plants and animals. Plant consumption varies seasonally and includes berries, tubers, forbs, and nuts. In Montana, John Waller and Richard Mace found that avalanche chutes provide rich forage for grizzly bears, due to the new plant communities that arise there. Grizzly bears prey on ungulates opportunistically, mostly on solitary moose (*Alces alces*) and on moose and elk (*Cervus elaphus*) calves. They also scavenge wolf kills and seek out gut piles left by human hunters.[8]

Grizzly bear diets also vary geographically, based on resources. An average grizzly bear diet consists of 10 percent mammal meat, 5 percent fish meat, and 80 percent vegetation. However, coastal brown bears have a considerably higher meat content in their diet, due to access to salmon (*Oncorhynchus* spp.), a high-fat, nutrient-rich food. In Yellowstone, which lacks sufficient berry-producing shrubs for bears due to twentieth-century over-browsing by elk, grizzly bear diets contain more meat than elsewhere in the northern Rockies. Primary Yellowstone foods include spawning cutthroat trout (*Oncorhynchus clarki*), ungulate calves, whitebark pine nuts, and army cutworm moths (*Euxoa auxiliaris*), foods subject to wide annual fluctuations in their availability.[9]

In Yellowstone, and probably elsewhere, grizzly bears are generalists with a broad diet. Their searches for alternative foods often lead to an increase in bear–human conflicts and human-caused bear mortality. Specifically, the decline in cutthroat trout, due to an illegal introduction of nonnative lake trout (*Salvelinus namaycush*) in Yellowstone Lake, along with whitebark pine decline due to a combination of the 1988 fires, climate change, and blister rust, have caused some bears to look for other foods in places where humans are present (e.g., campgrounds, people's homes outside the park). However, as we've seen, bears can eat over 200 different types of food, so they have many other food options and places to forage besides where there are people.[10]

Grizzly bears usually hibernate in dens once snow mantles the ground. In the northern Rockies, they den in the montane and subalpine zones, above the valley floor, favoring canyons and recessed areas in the landscape where there are few people. They don't eat or urinate during hibernation and burn fat to replace moisture lost through breathing. As the climate warms and there's less snow, grizzly bears may be entering hibernation later. Awareness of such patterns can help inform our decisions about grizzly conservation.

Before hibernating, grizzly bears gain up to several hundred pounds during a feeding frenzy called *hyperphagia*. One October in Glacier National Park, I observed a large male feeding in a Columbian ground squirrel (*Urocitellus columbianus*) colony. The bear went through the colony systematically, digging up more than a hundred squirrels during a two-hour period, popping them into his mouth whole and crushing them with his powerful jaws. Intense feeding and fat storage are essential to survive winter; a hibernating grizzly bear uses up to 4,000 calories daily just staying warm. Due to such high caloric needs, a pregnant female grizzly bear going into hibernation with less than 20 percent body fat will reabsorb her fertilized ovum before it implants, because her body will be unable to sustain a pregnancy.[11]

Conservation biologists have been working hard to identify grizzly bear habitat needs. They've found that grizzlies thrive where they have abundant food, low human presence, and corridors that enable them to move from one grizzly population to another. Formerly creatures of the plains and mountains in the western United States, in the 1900s, grizzly bears left the plains in order to avoid people. Mountain grizzlies were what remained after extirpation of plains grizzlies by humans. Thus, today, optimal grizzly bear habitat lies in remote mountain regions. Human development has created a gap in suitable habitat between Yellowstone and the Bob Marshall Wilderness to the north. Consequently, in recent years the growing grizzly bear population has been spilling east from the mountains, where grizzly bears can depredate (e.g., kill livestock). However, some individuals, such as a female grizzly that Mace radio-collared, spend their adult lives on the prairie, living peacefully in ranch country with their cubs.[12]

Ecological Effects of Grizzly Bears

In considering the big picture of large carnivore conservation, it's important to acknowledge each species' "job," or ecological role. Grizzlies don't drive top-down trophic cascades, due to their omnivory and hibernation. They eat many things besides meat, which weakens their direct impact on ungulates. Additionally, sleeping four to five months out of the year reduces the amount of time they spend eating plants and animals.

However, grizzly bear presence can enrich ecosystems in other ways. For example, grizzly bears scavenge remnants of wolf-killed prey, sometimes even chasing wolves off carcasses before the wolves have finished eating. This causes wolves to have to kill more often. Wolf biologist Rolf Peterson found that where there are bears, wolves have stronger top-down effects.[13] Additionally, by turning over the soil with their fork-like claws, grizzly bears give a boost to plant communities. In Glacier National Park, scientists found that glacier lilies (*Erythronium grandiflorum*) establish themselves best on subalpine meadows in the disturbed bare soil left behind after grizzly bears dig. This creates a positive feedback loop for lilies *and* bears, which eat this plant.[14] In Alaska, brown bears feeding on salmon along coastal streams increase inland soil nitrogen (N) via midden heaps (places where they deposit the remains of the salmon they eat) and via defecation and urination. Bears further enhance N transfer by disturbing the ground as they walk and by digging with their long claws, in effect aerating the soil.[15]

Grizzly Bear Conservation History

Grizzly bear conservation resembles the other large carnivores' stories. While humans have always hunted carnivores in moderation, modern humans have increased hunting's impact on wildlife communities by engaging in *market hunting* (i.e., hunting animals in volume to obtain meat or other commodities) and *trophy hunting* (i.e., hunting not for food, but to obtain specimens of male animals that set records for their body and antler size). Two thousand years ago, the emerging modern European culture saw carnivores as big, scary creatures that competed with humans and needed to be controlled. Two hundred years ago, European settlers brought these fears with them to western North America. Prior to that, grizzly bears ranged from Alaska to central Mexico, from the inland plains to the Pacific Coast. In California, their large population earned them a place on the state flag, although fossil records indicate that they didn't range as far south as Baja California. Ecologists David Mattson and Troy Merrill hypothesize that before European settlement, food was the primary regulator of grizzly bear numbers. Their distribution was lower in arid areas and highest in places with high-energy foods, such as acorns and pine nuts.[16]

Grizzly bear extirpation in the contiguous United States occurred mostly between 1850 and 1920. By the late nineteenth century, European settlers had killed almost all the bison (*Bison bison*), a key grizzly bear food on the prairie, and replaced them with carefully guarded livestock. Unsurprisingly, this had profoundly negative effects on grizzly bears. Then as now, many grizzly bears met death at the hands of humans over livestock depredation. Between 1920 and 1970, grizzlies continued to decline due to the growing human population and to drought. In Yellowstone, Mattson and Merrill linked a reduced *carrying capacity* (the number of individuals that can be supported by the habitat) for grizzly bears to climate change.[17]

In 1975, USFWS put the grizzly bear on the federal list of threatened and endangered plants and animals, under the Endangered Species Act (ESA). The legal term Congress uses for this action is *listing*. Congress uses the term *delisting* for the converse—removal from this list. The proper process for this calls for USFWS to propose listing or delisting a species by posting a notice in the *Federal Register*. "The list of federal threatened and endangered plants and animals" is recorded and kept updated in the *Federal Register*. By 1975 the grizzly bear inhabited less than 2 percent of its former range south of Canada and occurred in six small, discrete populations, totaling 800–1,000 individuals. The 1982 Grizzly Bear Recovery Plan identified regions connected to Canadian grizzly bear populations, which include Montana's Northern Continental Divide Ecosystem, the Yaak/Cabinet region, and the Selkirk and North Cascade Mountains. Additionally, the Greater Yellowstone Ecosystem (GYE), which comprises the park and surrounding areas in Idaho, Montana, and Wyoming, became a focal area for grizzly bear conservation. As of 2013, 1,400–1,700 grizzly bears were living below the US-Canada border. Figure 4.1 depicts grizzly bear distribution in 2013 within the Carnivore Way.[18]

Canadian grizzly bear conservation also began recently. In the 1950s, Alberta outfitter and rancher Andy Russell was the first to recognize that grizzly bears should be conserved and that humans and bears can coexist peacefully, even in ranching communities. Thanks to Russell's efforts, which included outreach via filmmaking and books, and informed by the work of Aldo Leopold and Farley Mowat, Canadians began to shift from killing grizzly bears to learning to live with them. In 1980, bear attacks on humans inspired cleaner garbage management in the Canadian national parks, and by the mid-1980s, grizzly bears had gained protection there.[19] In 2002, the Alberta Endangered Species Conservation Committee recommended "threatened" status for the grizzly bear. In British Columbia, the provincial government Blue-listed the grizzly, designating it a "species of special concern." The Canadian federal government also listed the northwest grizzly, with a population of 30,000, as a "species of

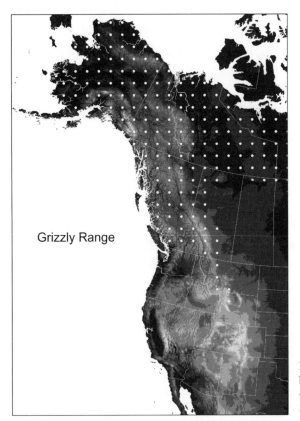

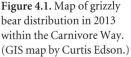

Grizzly Range

Figure 4.1. Map of grizzly
bear distribution in 2013
within the Carnivore Way.
(GIS map by Curtis Edson.)

special concern." But it would take until 2006 for the Alberta government to
suspend grizzly bear hunting and until 2010 to classify it as a "species at risk." As
of 2013 in British Columbia, grizzly bears had a population of 16,000 individuals, and they occupied 90 percent of their historical range. Alberta, on the other
hand, had 700 bears. No data has been published on the percentage of their historical range that they occupy in Alberta, although research has been underway
on bear distribution. According to Banff National Park wildlife–human conflict
specialist Steve Michel, "We can think of grizzly bears as the canary in the coal
mine. If they aren't doing well, it's an indication that Alberta's environment
might be compromised."[20] Grizzlies can be thought of as indicators because of
their low ecological resilience (specifically, their low reproductive rate). They'll
be among the first species to decline if habitat quality declines.

Are Bears Dangerous to Humans?

Human fear has presented a major threat to grizzly bear survival. Certainly a creature that weighs several hundred pounds and has long, sharp teeth and claws can inspire dread. However, most experts and people who live in grizzly country find bear–human relations to be fairly peaceful. A look at the actual risk grizzlies present to humans can help us move beyond fear.

Since 1900, close to 100 fatal grizzly bear attacks occurred in North America. Of these, 25 were in national parks. Given that over 60,000 grizzly bears live in North America, fatal bear attacks are very rare. Yellowstone bear management biologist Kerry Gunther puts this into perspective. "We average one bear-inflicted human injury per year. We have 3.6 million human visitors. More people commit suicide in this park than are injured by bears."[21]

As I mentioned in the opening of this chapter, one of the more recent fatal bear attacks occurred in Denali National Park. This was the first death of a human resulting from a human–bear encounter in that park's history. Forty-nine-year-old Richard White of San Diego had a passion for backpacking solo in wild, rugged landscapes. During his August 2012 visit to the park, he photographed a grizzly bear for eight minutes at close range (60 yards), and then was fatally mauled. The bear dragged White's body 150 yards into some willows and fed on him. A few hours later, three hikers came upon evidence of the mauling—blood, torn clothes, an abandoned backpack. They left the area immediately and reported what they'd seen. This park's bear-management plan calls for killing any bear that identifies humans as a food source. Rangers shot the bear the next day. A forensic investigation revealed White's remains in the bear's digestive tract. The images on White's camera showed the bear feeding calmly on shrubs before the attack.

White had received the mandatory backcountry safety orientation that all Denali overnight backcountry users get. This included recommendations to, when not in a vehicle, keep 300 yards from a bear, not backpack alone, and carry pepper spray.[22] Further, he'd made a trip to Denali previously and would have received the same advice and warnings at that time as well. White, an educated individual with considerable backcountry experience, disregarded some of this advice. For example, he was alone and too close to the bear. These tragic deaths, both human and bear, illustrate one of the biggest challenges of carnivore conservation—getting people to make better choices about how they interact with wildlife.

Coexistence between humans and bears begins with respect. Eminent bear authority Stephen Herrero has studied bear–human conflicts for over 40 years. He defines respect as "understanding the nature of an animal, accepting that, and then acting accordingly for their safety and yours." He observes that today, while most people may not have gained that level of understanding, more are acknowledging bears' wild nature and are behaving appropriately.[23]

Conflicts between bears and humans often have to do with food. Experts such as Herrero and Michel identify two categories of animal behavior with regard to humans. *Habituation* refers to an animal's high tolerance for humans, and *food conditioning* refers an animal's becoming accustomed to eating human foods, such as garbage. Habituated bears can coexist quite well with people under many circumstances. However, food-conditioned bears are very dangerous and sometimes attack people.[24]

Herrero and other experts recommend some basic bear-safety strategies. These include carrying pepper spray, traveling in groups, making noise, being alert, learning about bear habitat and bear behavior, and managing attractants (e.g., food, garbage). While appropriate in any context, these aren't one-size-fits-all prescriptions. According to Kevin Van Tighem, who writes about bears and has considerable bear experience, even though trends and patterns in bear behavior are very strong, no two bears will respond exactly the same in every situation, because, as is the case with humans, each bear is an individual.[25]

Canadian naturalist Charlie Russell, inspired by his father Andy, has devoted his life to studying bear behavior. Charlie has lived with human-habituated bears in Kamchatka, Russia, and coastal British Columbia. On his family ranch, which borders Waterton Lakes National Park, he and his family have coexisted peacefully with bears for decades. He believes that contemporary human conflicts with bears arose from the flawed paradigm used to manage our relationships with them. He explains, "Two specific ideas troubled me. One was that bears are unpredictable. In any conflict, if anybody got hurt by a bear, it was because the bear was unpredictable. The second was that if bears ever lost fear of humans, they were apparently dangerous right away. Managers saw fear as an important ingredient between humans and bears. And it needed to work both ways. Managers wanted people to be afraid of bears and bears to be afraid of people."[26]

Recently, this paradigm has been shifting. In Banff, Michel says, "People are seeing bears on the landscape for longer periods. And that's providing the opportunity for people to learn not only about the species, but about individual animals and family groups. People are recognizing that not all bear–human interactions are going to result in a dangerous outcome or will require a fear response. I always tell people that they shouldn't fear bears, but that they should have a healthy respect for them." Nobody really knows why bears, mostly females with cubs, are spending more time near humans. Some ecologists suggest that this may have to do with increasing numbers of male bears, which females with cubs consider to be a threat. Some think it may be related to climate change and shifting plant communities (e.g., decline of whitebark pine).

In Yellowstone, Gunther explains, "The foundation of our program isn't fear; it's preventing bears from obtaining anthropogenic foods and attractants." In Glacier National Park, there have been several fatal bear attacks, mostly related to garbage in the 1970s. However, this park hasn't had a bear-caused human fatality since 1998, or significant human injuries caused by bears since 2005. According to Glacier National Park ecologist John Waller, "We reach millions of visitors through our publications, educational messaging, and interpretive programs. Despite having millions of visitors and tens of thousands of backcountry camp nights every year, our visitor injury rates, property damage, and bear removals are astoundingly low and continue to decline."[27]

I've had many encounters with bears in the field. In learning to keep my family and field crews safe, I've consulted several bear experts. They've all advised me to follow the basic advice recommended by Herrero. Some have suggested that I speak to bears politely, the way I'd want a stranger, unexpectedly entering my home, to speak to me as he or she explained the intrusion. This advice has stood up well for me over the years. Others have tried this with similar results.

In 1976, while doing a survey in Glacier National Park, natural-resources specialist Clyde Fauley crested the brow of a hill and suddenly encountered a grizzly female with two cubs. The bears, who were approximately 150 feet ahead, immediately charged toward him for 50 feet, and then stopped, stood up on their hind legs, and sniffed the air. They growled, dropped to the ground, and charged again, this time stopping about 35 feet from him. Others might have panicked at this point. Fauley assessed the situation and decided against running for a large tree twenty feet away. Instead he spoke to the bears in a low, calm voice, saying, "Okay, now, just take it easy, bears, this is a no big deal—us park guys are on your side." As he eased his way back toward the trail and the temporary safety of a patrol cabin, he began to recite the park's bear management plan to them. He then spotted another large grizzly across the creek from the cabin, eating berries. The female and cubs were still in the area, the cubs standing on their hind legs to get a better look at Fauley. He arrived at the cabin, and then gradually worked his way from the cabin and back down the trail, still reciting park bear policy to the bruins. In the official incident report, he wrote, "My conversation with these bears would probably have sounded completely ridiculous to a second party. However, I believe it may have saved the day." He also noted that he never felt afraid. Had he reacted with fear and aggression, the outcome might have been different.[28]

Coexisting with Bears without Fear: Promises and Challenges

During Alaska's sockeye salmon (*O. nerka*) spawn, my companions and I travel from Denali to Katmai National Park and Preserve to witness brown bears feeding on these fish. Katmai had long been held up as a sterling example of peaceful coexistence between bears and humans. Then, one decade ago, two human deaths marred the relatively long history of good bear relations with humans here. Since those deaths, which we learn more about during our visit, Katmai bears have returned to business as usual, feeding on salmon while very close to people.

A DeHavilland Beaver floatplane carries us to Katmai. Our experienced pilot takes off on a southeast bearing from Anchorage, across the Gulf of Alaska toward the Alaska Peninsula, for the two-and-a-half-hour flight. Small clouds dapple the horizon on this otherwise clear day. As we gain altitude, in the distance I make out Denali, its majestic snowy top evanescent above the clouds. (While Denali is the correct name for this mountain, per the Alaska State Board, Mount McKinley is correct per the federal government.) Mount Redoubt rises next to us, a more solid form, glaciers at its feet. This is big country, utterly roadless, on a scale that's hard to absorb. Melting snow on the foothills creates a pattern of alternating green-and-white fingers of tundra and snow. This feels like untrammeled wilderness, yet gas rigs dot the Gulf of Alaska, a reminder that this place isn't quite as wild as it seems.

In Katmai, we descend into Brooks Camp over a green velvet landscape furrowed by rivulets and set down gently on Naknek Lake. The plane comes to a stop next to a half dozen other float planes on a narrow crescent beach. Dense alder and willow grow right up to the beach. The thick brush doesn't allow us to see what lies within the spruce forest that covers much of the land. Lots of bears, we know.

Congress created Katmai National Monument in 1918 to preserve the spectacular 40-square-mile ash flow deposited by the Novarupta volcano in 1912. A national park and preserve since 1980, Katmai is known today for its volcanoes, but also for the world's largest protected brown bear population (3,000 individuals) and its prodigious sockeye salmon run. Within Katmai, the Brooks River is a premier fly-fishing destination, and, on this river, Brooks Falls is the best place to observe grizzly bears as they fish for salmon. We're here at the salmon run's peak, which means we'll observe many bears.

We get out of the plane and our pilot leads us across the volcanic cobble beach to a well-worn forest trail. The forest feels eerie, claustrophobic. All at once, I pick up the earthy, pungent scent of wet fur. A bear. Very close—although I can't see it. A park ranger intercepts our group and turns us back, because a large bear is blocking the trail. We wait a few minutes for it to pass, and then continue. Human tracks comingle with bear tracks on the muddy trail, with bear scats the size of heaped dinner plates marking the way. The blackish-green scats smell of rotten fish—evidence of these bears' diet.

While I'm familiar with the danger presented by bears, I also know that it's quite possible to coexist peacefully with them, if one observes the etiquette of the wild. This means making no eye contact, not facing a bear squarely to minimize the threat you present, but at the same time not acting like a victim. A delicate balance, really. I hope these bears will be generous with us.

The Visitor Center compound consists of a small log building, a metal bear-proof shed, and a picnic area surrounded by an ominous triple-strand electric fence. Our pilot serves lunch at a table covered with a blue-gingham plastic tablecloth. Next, we receive a bear briefing. While I don't learn anything new, the context differs considerably from "down in the Lower 48" because the bear density is so high here. Yet even habituated bears have their limits.

Timothy Treadwell, a documentary filmmaker, devoted the last thirteen years of his life to the bears in Katmai. He came here in summer and filmed and interacted with them in order to advance bear conservation. To maximize contact, he set up his camp in the middle of bear travel corridors. He took none of the precautions recommended by experts, which included always having bear spray available and putting electric fencing around his camp. Treadwell reasoned that doing so would have breached his trust-based relationship with these bears.[29]

In October 2003, a bear mauled Treadwell and his partner, Amie Hugguenard, to death. The bear fed on them and reduced their bodies to shreds of flesh and bone shards. The hyperphagic bear had some health problems, which may have contributed to the attacks. Treadwell and Hugguenard's deaths are two of the three fatal bear attacks in this park in the nearly 100 years since its founding. Given the high bear density, this suggests that these animals have reached a rapprochement with humans unparalleled in most places.

Our briefing complete, we travel through the woods a short distance to a suspended bridge over the Brooks River. A park ranger reminds us that bears have the right of way. As she says this, we spot a bear walking along the lakeshore, about 200 yards from us. The ranger indicates that we can cross. As we do so, we notice a second bear in the water, about 50 yards from us and 20 feet from a fly-fisherman. We wonder about the fisherman's sanity, to be fishing here. And then we realize that the bears are everywhere and all around us—pale amber, just like in Denali, but much larger.

Upon crossing the bridge, we walk half a mile on a wide trail to a viewing platform. Dozens of heavily used trails riddle the woods, worn into the earth by the plantigrade feet of hundreds of bears over many years. The forest looks archetypical, like something out of Goldilocks and the Three Bears. We don't see more bears along the way, just lots of fetid green scats.

We hear the falls before we see them. They're wide, but not very high, banked by low-lying taiga on one side and forested, mossy cliffs on the other. The falls carry a tremendous volume of water—and fish. Sockeye are making their annual upstream run, their bodies flashing as they leap up the falls. A golden eagle (*Aquila chrusaetos*) perches on a cliff, eating salmon. But we're not really here to look at eagles. The fishing bears transfix us. I've never experienced anything like this spectacle.

Massive, 800- to 1,000-pound males convene here. Raw-boned and heavily muscled, they stand easily in the roar of the falls. Many have battle scars inflicted by other bears—gaping wounds, flaps of bloody skin hanging from their shoulders and flanks. Now and then they look toward us, their deep-set brown eyes incongruously small in their large, round heads. Their elaborate body language (standing in profile, posturing to make themselves look bigger, avoiding eye contact with other bears) conveys their primacy, yet avoids direct aggression. We don't see any females with cubs—this isn't their turf. The posturing between males continues during our entire time here. Occasionally a salmon leaps out of the water, swimming up Brooks Falls, and occasionally we see a bear catch one and then go through the elaborate ritual of eating it. Still standing in the full force of the falls, the bear grasps the fish in paws armed with impossibly long, sharp claws and slowly opens it, eating the soft flesh first, almost delicately, followed by the head (fig. 4.2). Sometimes the bears face into the current; extend and cup their large, wide tongues; and lap water as it comes off the falls. And sometimes they shake themselves off like big dogs, spraying water everywhere.

After a three-hour immersion in ursine life in all its drama, we start back to our plane, only to have a large, tawny bear cut us off. It strides by briskly,

Figure 4.2. Brown bear fishing in Brooks Falls, Katmai National Park, Alaska. (Photo by Cristina Eisenberg.)

looking straight ahead, as if we don't exist. However, as it passes, it swivels its round ears our way, tacitly acknowledging our presence. I observe that its front claws are much longer and more sharply curved than its rear claws—a detail I've never been close enough to a wild brown bear to observe until now. The better to eat salmon with, I suppose.

That night, I do some research and discover a controversy about hunting brown bears in Katmain National Preserve, which is contiguous to Katmai National Park. In Alaska, subsistence hunting and fishing are economically and culturally important. The Alaska National Interest Lands Conservation Act (ANILCA), a US federal law passed in 1980, gives rural residents the right to engage in subsistence practices on public lands near where they reside, including national preserves. Additionally, out-of-state hunters can pay up to $20,000 to guides for the privilege of hunting trophy brown bears on state lands. Considerable bear hunting occurs in the Katmai National Preserve. Such hunting raises an ethical question, because these human-habituated bruins allow people to get very near. This makes killing them too easy and violates the *fair chase* principle: wild animals should be able to range freely so that hunters pursuing them have no improper advantage.[30]

Meanwhile, 1,500 miles south as the crow flies, in Banff National Park, Alberta, a bear known as the Bow Valley Matriarch is teaching humans lessons of her own about coexistence. Officially called bear 64, this famous 25-year-old grizzly bear was initially radio-collared in June 1999, at the age of ten. She didn't have cubs in 1999 or 2000. Park wildlife managers don't really know what she did between 2001 and 2006, as she wasn't monitored during those years. They suspect, based on observations of other grizzly family groups within her known home range, that she had at least one litter during that timeframe. In 2007, managers saw her with two female cubs, which later became known as bears 108 and 109. DNA analysis of hair snagged at wildlife crossings showed that she actually had three cubs in 2006 (all female), but by the time people observed her in 2007 with her yearling cubs, only two had survived. Managers don't know the other cub's cause of death.[31]

Bear 64 and her two cubs became Banff favorites. Over the next three years, the cubs stayed at her side as she ranged within the Bow Valley, separating from her in 2009. Not long after that, tragedy struck. In 2010, a train hit bear 109; the next year, a car hit bear 108. According to Van Tighem, with an estimated population of fewer than 700 grizzly bears in Alberta and 60 in Banff, these deaths represented a significant ecological loss. Females are especially important to the persistence of a species with such a low reproductive rate. According to Michel, after a bumper buffalo berry (*Shepherdia canadensis*) crop that fall, bear 64 went into her den the fattest he'd ever seen her.[32]

In 2011, bear 64 continued to use the Banff town site. She came out of hibernation still in good shape, with three beautiful new cubs (fig. 4.3). In spring, she became an expert at hunting elk calves hidden in the tall grass on the edge of town. In early summer, she feasted on roadside vegetation, and in midsummer, she foraged at higher elevations. At summer's end, she returned to the valley floor to feed on buffalo berries. Remote-camera images showed her and her cubs using the Highway 1 overpasses and underpasses. Those images turned her into a poster bear for corridor ecology. Throughout that summer and fall, her

Figure 4.3. Bear 64 and her three cubs in Banff National Park, Alberta. (Photo by Amar Athwal.)

condition declined while nursing this litter, due to a poor berry crop. In spring 2012, she had her three cubs with her. That summer researchers fitted her with a GPS collar for a project aimed at preventing grizzly bear deaths on the railway tracks. Since then, her collar data have provided fascinating insights into how she uses the landscape—feeding on a variety of wild foods, ranging close to humans, but staying out of trouble. And in spring 2013, she emerged from hibernation with her three cubs still at her side.[33]

For most of the past 50 years, Parks Canada policy would have called for hazing bear 64 with rubber bullets to keep her farther from humans, or even for killing her, but these days, managers are trying a new approach: one of respect and peaceful coexistence. Bears like her make it easy to use such an approach. Michel points out that, while tolerant of humans and their infrastructure, bear 64 does a really good job of maintaining a necessary distance from people. His greatest concerns are the ongoing mortality threats she faces in the Bow Valley and on the railroad tracks.[34] In a perfect world, bear 64 would continue to raise healthy cubs and would eventually die of old age. Time will tell how she lives the rest of her years. But this wise bear provides living testimony of respectful relationships between humans and bears.

From Human Habituation to Food Conditioning: The Oldman Lake Grizzly

What happens when bears cross behavioral boundaries? And what are the challenges of reinforcing those boundaries? Approximately 175 grizzly bears make Glacier National Park home. Park staff work proactively with visitors to keep these bears alive. Strategies include educating visitors about bear safety and behavior, and managing daily interactions between people and bears to prevent the cascade of circumstances that can lead to food conditioning and death. However, in this park, as in all others, cost drives bear management. According to John Waller, sometimes bears that the park has worked very hard to keep alive are success stories in the short term, but end up dying at the hands of humans. While it's easy to focus on such failures, it's important to acknowledge the successes that occur in managing our daily human relationships with bears.[35]

Glacier has had many bear-management successes since the 1990s, situations where park staff and managers have worked proactively with grizzly bears and people to foster mutual respect and coexistence. However, when I ask Waller to share one of the most meaningful lessons he's received from a bear, he brings up the Oldman Lake Grizzly. He and other park staff put enormous effort into keeping this seventeen-year-old female grizzly bear and her cubs alive. But, unlike bear 64 in Banff, the Oldman Lake Grizzly didn't understand boundaries. She earned her name because of the popular backcountry campground where she often hung out, despite abundant huckleberries (*Vaccinium* spp.) and other foods in the surrounding mountains. While she never behaved aggressively toward humans—never huffed at or bluff-charged anyone—she insisted on getting too close. How close? She'd greet people on the trail into camp, investigate what they were cooking, and sniff their tents in the night.

In 2005, the park contracted Carrie Hunt of the Wind River Bear Institute to bring in her Karelian bear dogs to teach the Oldman Lake Grizzly the meaning of "No." Bred in Finland and Russia to hunt bears, these dogs are now used for *aversive conditioning* (i.e., linking undesirable behavior, such as spending time in campgrounds, with an unpleasant or noxious stimulus, such as the dogs). Karelians are friendly, mild-mannered dogs until they smell a bear, whereupon they instinctively spring into action, tracking and relentlessly harassing the bear until it leaves the area.

Hunt and her dogs worked with the Oldman Lake Grizzly in 2005 and 2006. They didn't just haze the bear, they tried to teach her more appropriate behavior. The park invested significant funds and effort into this process, which initially seemed to have worked. During the next two years, the rehabilitated bear was left in peace to integrate what she'd learned. She spent time in the backcountry and didn't approach humans. However, in 2008, she reverted to her old ways and resumed approaching people repeatedly. Unable to sustain the high cost of aversive conditioning, which didn't fully "take" with this bear, the park had to make the agonizing choice to destroy her.

Waller explains,

> I'd hoped that she could really teach us a lot about how bears and people might coexist, but in the end, she was destroyed. I still struggle to decide whether the outcome was a success or failure—I guess it was a little of both. A failure in that we didn't learn what I'd hoped; a failure in that she was destroyed. But on the other hand, she survived long enough to rear several sets of cubs. We did learn quite a bit about how to teach bears. And her destruction moved the park to reaffirm its commitment to bears and bear conservation.[36]

Recovery of the Great Bear

In the 1970s, grizzly bears in the contiguous United States were far more at risk than those in the far north and Canada. This resulted, in 1975, in federal protection of grizzly bears under the ESA. In 1981, USFWS appointed Christopher Servheen as Grizzly Bear Recovery Coordinator. By 1983, Congress had approved a recovery plan and convened the Interagency Grizzly Bear Committee.

Between 1975 and 2007, the number of Yellowstone's grizzlies increased from 136 to 571. Given this progress, in 2008, the federal government delisted the species.[37] The Greater Yellowstone Coalition litigated the delisting, citing whitebark pine decline and inadequate regulatory mechanisms to protect grizzlies post-delisting. US District Judge Donald Molloy ruled in favor of relisting the species due to USFWS's failure to apply best science in grizzly bear management. Environmental attorney Doug Honnold explains,

> Whenever the Service makes a listing determination, whether it's an action to put a species on the list or an action to take a species off the list, it has to use the best available science. So if it disregards a key scientific issue, like the loss of whitebark pine, then the courts are going to say that's not good enough, the species deserves better, and the ESA requires use of best science.[38]

From 2008 to 2012, grizzly bear numbers in the GYE held steady at over 600. Some experts, such as Gunther, believe this population has finally reached carrying capacity. He suggests that, because the grizzly is an omnivore, whitebark pine decline is unlikely to create a bear decline.[39] After the grizzly bear was relisted, Judge Molloy asked the Interagency Grizzly Bear Study Team, a group of state and federal officials, to look further into whitebark pine issues and grizzly bear demography.[40]

Grizzly bear demographic assessments are complicated by this species' low reproductive rate. Recovery depends on the portion of the population capable of reproducing, called the *ecologically effective* population. (This differs from how conservation biologists define this term to mean a species' ability to drive trophic cascades.) As population ecologists and USFWS use the term, species with a low reproductive rate will have a low ecologically effective population. For example, a population of 100 bears may have only 15 reproducing females. Scientists have identified a sustainable grizzly bear mortality for females with cubs less than a year old (called *cubs of the year*) of 4 percent. In 2013, USFWS adjusted this to 7.6 percent, based on new science.[41]

Determining how grizzlies are to be counted has emerged as one of the leading challenges to the federal grizzly bear delisting proposal. In 2013, US-

FWS found 718 grizzly bears in the GYE, well above the recovery threshold of 500. However, this count may have been biased because pine nut and cutthroat trout declines have altered grizzly bear feeding patterns, driving them to feed on alternative high-protein, high-fat foods such as army cutworm moths. The moths live on open, rocky slopes, above treeline. According to Daniel Doak and Kerry Cutler, bears eating army cutworm moths are more visible and easier to count, making it seem like bear numbers have increased. They suggest that the GYE grizzly population has probably increased far less than is believed. In 2013, in response to questions regarding habitat, climate change, and how one counts grizzly bears, USFWS published several recovery plan supplements.[42]

Since 1981, grizzly bear recovery has entailed an extensive transboundary, collaborative effort. Servheen has led the US portion of the Trans-Border Grizzly Bear Project since its inception. He explains, "We share grizzly bear populations with Canada. The bears have dual citizenship. And we work to maintain that connection, so that it doesn't disappear. We work very closely with our Canadian colleagues on issues of linkage, habitat management, and mortality. This close cooperation enhances our ability to make management decisions." Focal areas for this effort have included the Selkirks, Cabinet-Yaak, and North Cascades. According to Servheen, these recovery areas must be connected with Canadian grizzly bear populations in order to maintain healthy numbers in the United States. As we saw in chapter 1, ecologist Michael Proctor has identified the Highway 3 corridor as one of the more formidable movement barriers for most mammal species, particularly grizzly bears.[43]

In the United States, post-delisting, the states will be responsible for grizzly bear management. Each of the three states (Idaho, Montana, and Wyoming) in the GYE has a grizzly bear management plan. The plans call for establishing a primary conservation area (PCA) within the tri-state region and managing toward an overall population goal. The federal recovery plan calls for maintaining 500 grizzly bears within the PCA, with at least 48 breeding females. All states will allow hunting of grizzly bears, although not immediately after delisting.

But recovery isn't just about how many bears exist. According to Servheen, the long-term future of grizzly bears "is going to be based on the support of the people that live, work, and recreate in grizzly bear habitat. Our challenge is maintaining that stakeholder support." To be effective, grizzly bear conservation must take place at the grassroots level. The potential for conflict on ranchlands and leased grazing allotments makes such an approach crucial. In the GYE between 1992 and 2000, 44 percent of bear–human conflicts had to do with livestock depredation. Depredation peaked prior to hibernation and occurred independently of wild-food availability.[44] Montana Fish, Wildlife, and Parks grizzly bear management biologists Tim Manley and Mike Madel work tirelessly to help ranchers, hunters, and bears coexist. They arrive on the scene when there's a problem, patiently answer questions, and offer suggestions for making homes and ranches more bear-proof.

Conservation biologist Susan Clark maintains that conflict resolution depends on consensus. Montana's Blackfoot Challenge provides an inspiring example of how consensus building can help keep grizzly bears alive. In the 1970s, this group of rural neighbors along the Blackfoot River began to collaborate with each other and with agencies to create healthier working lands—a radical approach in a state known for rugged individualism. Their approach was so effective that in 1993 these landowners and partners formally formed the Blackfoot Challenge. Since then they've become widely known for stewardship practices such as their range-rider program, in which people on horseback ride among herds of cattle to prevent depredation.[45]

In Alberta, protecting grizzly bears within the national parks caused their numbers to increase outside the parks as the bears dispersed onto ranchlands, necessitating community-based conservation. Gibeau and Herrero work with the Nature Conservancy Canada proactively to reduce conflicts between bears and ranchers. They begin by developing trust-based relationships with ranchers and then help them apply stewardship practices such as storing grain in bear-proof containers and removing livestock carcasses promptly.

In the 2000s, Alberta Environment and Sustainable Natural Resources Development added a complementary program involving roadkill. In the spring, when grizzlies emerge from hibernation ravenously hungry at the same time that cattle are calving, managers use helicopters to drop road-killed deer and elk carcasses in the foothills on public lands. Called the Intercept Feeding Program, these carcass drops draw bears like magnets and may be helping reduce depredation on livestock. Some of the drop sites lie in Waterton Lakes National Park on spruce-clad montane benchlands high above valleys. Ecologist Barb Johnston has been monitoring these drop sites with trail cameras and is finding grizzly bears feeding there along with other large carnivores. Such efforts are demonstrating that with ingenuity, collaboration, and commitment, it's quite possible to improve coexistence between people and grizzly bears.

CৡৡৢO

Our tangled history with the great bear ranges from forming a close spiritual bond in the era before the European settlement of North America to reviling this species as a threat to our livelihood. The real grizzly lies somewhere in between. This powerful carnivore has the potential to do great damage—but usually doesn't. If treated with respect, it has the capacity to coexist with us peacefully. Its presence enriches landscapes ecologically and intangibly: a landscape that contains grizzly bears feels far wilder than one that doesn't. Social scientists maintain that such wildness is part of what makes ecosystems and humans whole. According to ecopsychologist Peter Kahn, "Today wildness remains part of the architecture of the human mind and body, and to thrive as individuals and as a species, we need to cohabitate with it."[46] The bears and people profiled in this chapter eloquently tell the story of our struggles to renew the great bear's presence in our lives. In the next chapter, we'll see that wolves are teaching us similar lessons about coexistence.

Wolf (*Canis lupus*). (Drawing by Lael Gray.)

CHAPTER FIVE

Wolf (*Canis lupus*)

In the summer of 2002, when my young daughters and I saw a pair of wolves (*Canis lupus*) tear across the meadow on our land chasing a white-tailed deer (*Odocoileus virginianus*), we tracked them. To avoid interfering with their hunt, we waited a few minutes before following the swath of flattened grass they'd left across our meadow. Their trail led us into the forest, to an old barbed wire fence that bounds our property. The deer had gone over the fence and onto our neighbor's land, and the wolves had followed. But rather than jump the fence, the wolves had run between two strands of wire, leaving clumps of hair on the barbs. I collected it, went indoors, and glassed it with a hand lens. The samples had a thick, oily undercoat and long, dark guard hairs.

The next day, my daughters and I went to our nearest natural-resources office to report what we'd seen. The twenty-something person staffing the US Forest Service front desk smiled politely and said, "Those were probably just big dogs." Back then, there were no known wolves officially confirmed in the area where we lived.

I didn't give up. The next week, I made an appointment to meet with Tom Meier, biologist with the US Fish and Wildlife Service (USFWS or "The

Service") and the federal Wolf Recovery project coordinator for western Montana. This was the same Tom Meier whom I would visit years later, in Denali National Park. Back in 2002, when I sat in his small, cluttered Kalispell office and told him about the animals I'd seen, he said "Oh, really?" and raised an eyebrow. I handed him the clump of hair from our fence, which I'd put in a Ziploc bag. He examined it closely for some moments. Then he looked up and asked, "Wanna track 'em for me?"

By that winter, I was tracking wolves for Tom from my writing cabin, which lay at the locus point of three wolf packs' territories. Two of the packs had recently formed, and Tom knew little about them. I'd go out on snowshoes or skis and find their trails as they hunted elk. Sometimes I'd find the tracks of a half dozen adults and their nearly grown pups. I could tell the generations apart, because the adults left elegant, energy-conserving trails that vectored through the landscape, seldom veering, while the curious pups left meandering trails as they checked out rocks, trees, fences, sticks, and anything that moved.

The next year, at Tom's suggestion, I began a master's degree on wolf ecology and recovery, and he left to work in Denali. I didn't see him again for nine years, until after I finished my PhD and went to visit him, excited about learning more about wolves in the far north.

The Wolf Townships

Tom began studying wolves in Denali National Park in the mid-1980s, on a project led by David Mech. Over a ten-year period, they and co-researchers Layne Adams, John Burch, and Bruce Dale looked closely at wolf movements and relationships with prey. Their radio-collar data depicted the fascinating "boom and bust" economy of wolf life in the far north, in which prey availability affected everything from pack territory size and social dynamics to whole food webs.[1] When that program ended, Tom went to work for USFWS in northwest Montana. In 2003, he returned to Denali—a place he loved.

In 2012, I visited Tom in his comfortable, light-filled park office, and when I asked about wolf ecology and conservation in Denali, he was off, telling stories in his soft Minnesota accent. He told me about wolves killing twenty caribou (*Rangifer tarandus*) one hard winter, and then shaving the meat off the frozen carcasses for weeks, even denning near them. He explained that in addition to prey availability, trapping by humans outside the park affected the Denali wolf population.

A large Denali map hung on Tom's wall. He pointed at the northeast corner of the park, where the map showed a notch seven miles high and twenty miles deep. When Congress expanded the park in 1980, they didn't include this chunk of private land, a popular hunting and trapping area called the Stampede Corridor and Wolf Townships. Tom described a recent incident there to illustrate the conservation challenges that arise in areas where wolves are subject to legal killing beyond park boundaries.

In May 2012, a trapper hauled a dead horse to a riverbank in the Wolf Townships near the park boundary. He set snares all around it, hoping to catch wolves attracted by the carcass. The snares lay in an area formerly managed as a buffer zone, where for one decade the state had prohibited wolf trapping in order to protect wolves that spend a good portion of their lives inside park boundaries. The state had lifted this restriction in 2010, with no plans to reinstate it. The trapper caught two wolves. One was a breeding female of a pack often seen by park visitors—the Grant Creek pack. To make matters worse, in June, the other breeding female in this same pack died of natural causes. Thus it appeared there'd be no pups in the pack. The *necropsy* (wildlife autopsy) Tom conducted on the first dead wolf revealed that she'd died from being in the trap for a prolonged period. However, a wolverine had scavenged the carcass, making it difficult to determine her actual cause of death. While the trapper had done nothing illegal, this wolf's death raised public ire and prompted an emergency petition to reinstate the buffer, which the state denied.[2]

According to Yellowstone National Park Wolf Project leader Doug Smith, at the heart of such transboundary issues lies the fact that the US National Park Service (NPS) has a mission to *preserve and protect* natural resources, which means no consumptive or destructive use (parks are intended to be used for public enjoyment). Meanwhile, states have a mandate to *conserve* natural resources, which means wise use and can include hunting. In places like Denali and surrounding lands, these two different mandates collide both philosophically and practically.[3]

CB&D

The next day, my traveling companions and I joined Tom and park wildlife biologist Bridget Borg on a carnivore survey focusing on wolves, and specifically, on the East Fork wolf den. Discovered by Adolph Murie in1940, this den site had been used periodically by wolves throughout the decades. In the late 1930s, administrators hired Murie to help resolve a wolf management issue in Denali, then called Mount McKinley National Park. The Campfire Club, a Washington, DC, hunting organization, demanded that the park kill its wolves in order to boost Dall's sheep (*Ovis dalli*) numbers and create a sanctuary for breeding so that there'd be excess animals to replenish adjacent areas for hunters and tourist viewing. The McKinley wolf controversy tested the new NPS carnivore-protection policy, established in 1934 by George Wright.[4]

We crossed a stream and then bushwhacked through the willows along the East Fork River. We paused periodically and listened for radio-collared wolves, scanning for signals using a receiver and Yaggi antenna (a hand-held antenna shaped like the letter "H"), but we heard nothing. Murie's East Fork den lay atop a high knoll. Ribbons of aspens (*Populus tremuloides*) grew on the knoll's south-facing flank. Leaving the riverbed, we side-hilled and scrambled up a steep, partially washed-out talus slope, rested briefly on a grassy sward, and then thrashed through hellacious shrub thickets toward the den, 200 yards farther up.

The den's dark, oval mouth lay in a slope of red-ochre sandy soil, topped by a thick thatch of grass and azure forget-me-nots (*Myosotis alpestris*). From this vantage point, we could see for miles along the East Fork River to Polychrome Mountain and beyond. In 2011, the Grant Creek pack had used a nearby den located in an aspen copse about one quarter mile from the Murie den. Tom and Bridget had observed the Grant Creek pack using that other den during spring and early summer of 2012. They surmised that since the demise of the pack's two breeding females, a third female may have bred.

Near Murie's East Fork den, we found at least a dozen other den holes. The holes had been cleared of vegetation by wolves, a sign that they'd been used earlier in the season. However, other than a clump of gray-colored wolf hair snagged on the fine grass roots at the Murie den entrance, we found no evidence of more recent wolf use.

We sat outside the Murie den and talked about wolf studies as we ate lunch. There are places where scientists have had insights that have profoundly changed how we see the natural world. This was one of them.

In the 1930s, we knew little about wolves. In his landmark study, Murie patiently pieced together the natural history of wolf social ecology and hunting habits. He spent weeks sitting on a knoll across the river from this den, observing it and its resident wolves through binoculars. Murie found that a pack of seven adults and five pups were using the den. He determined that they preyed mainly on sheep, primarily killing young and weak animals, and so he concluded that wolf predation had a salutary effect on the sheep population. This countered the thinking of those who favored culling park wolves in order to increase Dall's sheep numbers.

After World War II, he recommended continuing wolf control in the park until the sheep population showed signs of rebounding, a view popular with the critics of the NPS. Murie organized the wolf-control effort for the next few years in order to be specific about which wolves would be sacrificed for the politics of wildlife management; in particular, he sought to protect the East Fork pack. Over the next five years, Murie ended up removing a handful of wolves that would have died anyway (i.e., sick, old, and weak individuals). All along, McKinley Park opted to use Murie's science to manage its wolves, which meant continuing to protect them as much as was practically possible so they could serve their ecological role. The McKinley wolf-control program quietly ended by the early 1950s.[5]

Tom pointed out that the landscape before us probably hadn't changed much since Murie's time. The East Fork River still flowed broad, braided, and silty over extensive gravel bars. Talus slopes and high, round hills, such as the one we were on, framed the valley thus formed. And Polychrome Mountain still colorfully dominated the panorama.

Over the twenty years that Tom had worked here, the dynamic equilibrium between wolves and sheep—this trophic dance—had continued as in Murie's day. Tom explained that the wolf's preferred prey had shifted since then several times, due to changes in prey accessibility and climate. Currently, some packs preferred moose (*Alces alces*), followed by caribou, and Dall's sheep. However, in the eastern portion of the park, wolves such as those in the Grant Creek pack depended primarily on caribou. Further, these preferences also changed seasonally based on what was easiest to hunt. For example, in deep snow, caribou were easy for wolves to kill. And in addition to seasonal cycles in wolf predation, longer cycles occurred over time, driven by variations in the severity of winter.

Others followed Murie in studying these wolves. Gordon Haber conducted wolf *ethology* (behavior) studies here from the early 1970s until 2009. For several decades, Vic Van Vallenberghe studied Denali's moose–wolf interactions. And for her PhD, Bridget was looking at how wolf trapping outside the park affected wolf viewability—that is, the likelihood of wolves being seen by visitors inside the park.[6] Collectively this work built on the observations Murie had made so long ago with binoculars and a pencil and notebook. Yet the fundamental truths he found—that wolves were social creatures linked to their prey in a dynamic equilibrium that turned around the seasons and ebbed and flowed with climate and other ecological factors—underlay all the recent science.

We gazed out at the Toklat Valley and talked about geology and conservation. This big landscape bred big mountains and bigger thoughts. Its ineffable wildness had inspired science rooted in both empiricism and a deep love of nature. Tom was part of this legacy. We talked about hope. And as we discussed the vicissitudes of wolf management, he reminded us that, as with all else, we couldn't survive on sorrow and anger.

We got up and bushwhacked another 100 yards upslope, where we found still more dens. We had to walk carefully to avoid stepping into one accidentally, because the tall grass sometimes hid their openings. At the top of the knoll we found a meadow spangled with yellow cinquefoil (*Potentilla* sp.) and what Tom had been searching for: the carcass of the Grant Creek pack's second alpha female. She had died that spring of natural causes, perhaps while giving birth. In her lifetime, she had many pups, who filled this landscape with their howls and wildness. She lay on her side on a soft carpet of grass, her carcass intact and beautiful. Tom considered collecting her skull for a park specimen. He knelt, gently touched her thick, pure-white fur, and noted her well-worn teeth. Loath to disturb her, he decided to let her be.

What's in a Name: The Dog-Wolf

While in England giving talks about wolves for the UK Wolf Conservation Trust, I spent a rainy afternoon exploring London's Natural History Museum. I must confess, I went there for a thrill: to see the world-renowned Dinosaurs Hall, complete with a moving, roaring, life-sized T. rex replica. However, on the way to visit this king of the dinosaurs, I got sidetracked. As I passed the Mammals Hall, I saw a throng of people pressed against a bank of glass cases. Curious, I worked my way to the front of the mob. The cases held the museum's carnivore collection, which included the fossilized skeleton of a dire wolf (*Canis dirus*), an early wolf. A boy commented on the sharpness of its huge teeth. "The better to eat you with," said his friend, laughing. A woman softly said, "I wonder if it was as beautiful as today's wolves."

Nobody has figured out why wolves so captivate human imaginations. This creature stops people in their tracks, from accomplished biologists like Tom Meier to English schoolchildren to housewives like me with wolves running through their yards. To parse our storied relationship with this animal, which is similar to humans in some ways and perhaps embodies wildness more than any of the other carnivores, the best place to begin is with its taxonomy and natural history.

The wolf is a wild dog, the largest member of the mammalian family Canidae. The wolf genus *Canis* developed in North America in the late Miocene epoch, about 10 million years ago. This genus was the forerunner of all the wolves (*C. lupus*), coyotes (*C. latrans*), and domestic dogs (*C. familiaris*) we have today. By the early Pleistocene epoch, 1.5 million years ago, wolves had split off from coyotes and crossed into Eurasia via the Bering Land Bridge, where they continued to evolve into *C. lupus,* which means "dog-wolf."[7]

During the mid-Pleistocene epoch, about 700,000 years ago, successive waves of wolves returned to North America from Eurasia. Meanwhile, the dire wolf, *C. dirus* ("fearsome dog"), arose in North America 300,000 years ago and spread into South America. Stockier than today's wolf, it had a massive head, large teeth, and relatively short limbs.

Regional adaptations to habitat, prey, and climate caused the *C. lupus* subspecies to develop. Until recently, biologists recognized 24 subspecies of gray wolf in North America. However, since the 1990s, DNA studies have shown that far fewer exist. These subspecies reflect Pleistocene waves of wolf colonization. The first wave brought the Mexican gray wolf (*C. l. baileyi*) to southwestern North America; the second wave brought the plains wolf (*C. l. nubilus*); and the third brought the Alaskan wolf, also known as the timber wolf, (*C. l. occidentalis*), to the far north. The dire wolf became extinct for unknown reasons about 8,000 years ago as part of late-Pleistocene extinctions that included the saber-toothed tiger (*Smilodon*). The plains wolf, which became extinct in 1926, once occupied most of the western United States, southeastern Alaska, and central and northeastern Canada. Commonly called the "buffalo wolf," it may have had the greatest distribution of any of these subspecies. Today, within the Carnivore Way, only two wolf subspecies exist: *C. l. occidentalis* and *C. l. baileyi.*[8]

Wolf Natural History

The wolf is one of the most powerful and adaptable predators worldwide. Its natural history reveals why it's an apex predator. Per the scientific literature, adult males average 85 to 100 pounds, and adult females average 80 to 85 pounds. However, according to Doug Smith, wolf weights can vary greatly, and the actual range observed in the field may be wider than the above averages. Primarily gray, wolves also can be white or black, with much individual variation. Smaller and browner than the timber wolf, the Mexican gray wolf subspecies otherwise has similar traits and behavior. Superbly equipped for predation, the wolf has sharp *carnassial teeth* (molar and premolar teeth with jagged shearing edges) that efficiently slice through hide, flesh, and bone. Its bite packs more than 1,200 pounds per square inch (PSI) of crushing power (compared to 600 PSI for a large domestic dog). Its sense of smell is among the keenest in the animal world. The old Russian proverb, "The wolf is kept fed by his feet," holds much truth. Physically capable of traveling more than 40 miles per day, it can run 35 miles per hour and swim 50 miles. This adaptable species has two principal habitat needs: abundant ungulate prey and low conflict with humans.[9]

The fundamental unit of wolf society is the pack. Wild wolves live in packs averaging six to ten wolves, but pack sizes of up to 43 animals have been recorded. Packs are families in which adult parents shape group activities and share leadership with other wolves. Wolf families have a hierarchical structure, from the breeding (or *alpha*) male and female at the top to lower-ranking animals, although this hierarchy isn't as simple as scientists once thought. The breeding male and female are strong hunters and the most capable of maintaining pack cohesion.[10]

Each pack has a territory. Territory sizes vary, depending on habitat, prey, local wolf density, and human population. Territories can range from 20 to 2,700 square miles, with those in the far north generally larger than those in the south due to varying ungulate densities. In the northern Rocky Mountains, the average territory size varies from 100 to 200 square miles. Wolves form new territories by dispersing. A key survival strategy, dispersal helps wolves thrive because it leads to genetic exchange. In Montana, Diane Boyd found an average wolf dispersal age of three years, and an average distance of 48 miles for females and 70 miles for males.[11] However, wolves have been documented dispersing more than 1,000 miles. To be able to disperse, they need safe passages through landscapes and connected travel corridors.

The wolf has a higher reproductive and survival rate than many of the other large carnivores. Females first whelp pups at two to five years of age. Usually only the alpha pair breed, although occasionally more than one pair breeds in a pack, leading to multiple litters. Breeding season runs from late January through February in the northern Rocky Mountains, and then later the farther north you go and earlier in the south. Pregnant females dig dens, where, after a 63-day gestation, they give birth to four to six pups. Pups remain inside the den for their first four to six weeks of life. At four weeks, the alpha female begins to wean them. Their socialization takes place away from the den at resting areas called *rendezvous sites*. Pups develop rapidly, and by early winter they have nearly reached adult size. Generally 70 percent of pups survive their first year, although lower survival rates can occur due to disease (e.g., canine parvovirus, distemper).[12]

Hunting is risky business for wolves, which must have meat to survive. These *obligate carnivores* (defined as carnivores that can derive nourishment only from meat) select prey based on availability, vulnerability, and profitability of the hunt. This means they first focus on prey that is easiest to kill, switching to the species with greatest biomass when it's energetically advantageous to do so. For example, it may take as much energy for wolves to run down a deer (*Odocoileus* spp.) as an elk (*Cervus elaphus*), but an elk provides more meat. Because of this, where elk abound, they're wolves' first choice. And while bison (*Bison bison*) provide more meat than elk, south of the US-Canada border wolves tend to avoid them, because bison are more dangerous than elk. However, in Yellowstone, as elk numbers have dropped, wolves have been killing more bison. Regardless of the prey species, wolves give up the chase when the risk becomes too high. To further minimize risk, they kill at a rate that often keeps them minimally nourished.[13]

Most hunts are unsuccessful for wolves. Doug Smith found that in Yellowstone National Park only 5–15 percent of wolf hunts succeed. Portland filmmaker Vanessa Renwick has eloquently depicted these odds with her art installation *Hunting Requires Optimism*, in which she invites people to open each of ten refrigerators on display. However, instead of finding food, they discover television monitors with ten different video loops of wolves in a winter landscape stalking and only once actually taking down and eating prey.[14]

It's not so easy to be a wolf. Wolves have short lives because of all the hazards that they face. Smith has found that the median age of death of a wolf in Yellowstone National Park is five years (lower than the median age of death for bears or cougars).[15] The species makes up for the challenges wolves face by having a high reproductive rate and by being adaptable in their habitat and nutritional needs. This adaptability, plus their high physical endurance and reproductive rate, makes them very resilient.

The Ecological Benefits of Wolves: The Science Hunt

Scientists have thoroughly documented the wolf's keystone effects in places like Yellowstone National Park. However, these effects are becoming widely recognized in ecosystems beyond national parks, sometimes surprisingly so.

In December 2012, I spent a week hunkered down in the hoarfrost, hiding in the sage (*Artemisia tridentata*) and fording icy streams during a late-season Colorado cow-elk hunt on the High Lonesome Ranch, a 400-square-mile conservation property. As elsewhere, in this place with too many elk and not enough wolves, hunting cow elk provides a powerful conservation tool because of its effectiveness in thinning herds. My hunting companions included conservation biologists Jim Estes, Michael Soulé, and Howard Whiteman.

Day after day, we stalked elk. Invariably, the elk took off before we were within shooting range (e.g., 400 yards). One day, we bushwhacked up a mountain during a snowstorm, slogging through two feet of fresh snow, seeing elk tracks and elk everywhere, and still were unable to get within shooting range of these wary animals. We didn't care—it felt good to be outdoors, working hard. And if we didn't get an elk this year, we'd try again the next.

On the fifth day, we were bumping down a dirt road at midday in a ranch truck, planning our afternoon hunt, when we came upon a beat-up orange flatbed pickup headed toward us. The truck slowed and stopped. We pulled parallel to it to chat, as is the custom in the rural West. The driver, a local who'd lived here all his life and worked at a nearby gas well, asked if we'd had any luck. We told him no.

"Me neither," he said, "And I've been hunting for nearly a month. I've never seen these elk so wild!" Then his eyes got big. "It's the wolves!"

We looked at each other and said, "Wolves?"

"Yeah," he said, "everyone knows about them." He proceeded to describe a pack of wolves that frequently dipped down into Colorado from Wyoming. "They're making the elk wild again!"

At this point I should mention that all of us scientist hunters looked like a bunch of, well, hunters. We were wearing camouflage jackets, blaze-orange vests, and wool hats with ear flaps. After several days of hard bushwhacking, we looked much the worse for the wear. The hunter who stopped us had no way of knowing who we were, or that among us we'd written a dozen books and scores of journal articles on carnivores, food webs, viable populations, and conservation biology. We kept quiet about this and just listened to what he had to say. Quite simply, he was full of wisdom. He went on to give us a textbook-perfect explanation of trophic cascades and the ecology of fear.

"The elk never know where the wolves are going to be next," he said, "so they have to stay on the move. Man, I've never seen them so wary! They used to just stand around, eating everything to death. Now maybe they'll give aspen a break, and we'll have healthy forests again."

Fascinated, we asked about the elk when he was growing up. "They just stood around like lawn ornaments!" he said. "I sure would like to get some elk to feed my family. But on the other hand, it's cool to see them so wild!"

We spent the rest of the day doing what ecologists do—talking about all the other possible factors that could have been influencing elk behavior. We never did get our elk that year. But like our friend in the orange truck, we enjoyed seeing them so wild. This random meeting with a hunter demonstrates that while we have far to go with the human factor of wolf conservation, we're making progress.

Ideas about the powerful link between predators and their prey originated in the early 1900s. In addition to Adolph Murie's work, his brother Olaus and their colleague Aldo Leopold noted that throughout North America, where humans had removed wolves, ungulates were heavily browsing saplings and shrubs, to the point that few aspens were able to grow above the reach of animals like elk. In describing the devastation that he found, Aldo Leopold wrote, "I have seen every edible bush and seedling browsed, first to anemic desuetude and then to death."[16]

As we saw in chapter 2, predation, or even its threat, shapes how plant-eating animals feed. In a trophic cascade, an apex predator like the wolf indirectly benefits plants by both killing and scaring elk, thereby lessening the pressure elk put on food sources such as aspens. This enables aspens to grow into the forest canopy and makes habitat for other species, such as insects.[17] The patterns in Colorado, in which recolonizing wolves may have been making elk more wary, represent the first step in a trophic cascade.

Wolf Conservation History

Wolves along the Carnivore Way have a checkered past, both historically and more recently. Several hundred thousand wolves lived in North America before European colonization. In the 1880s, while giving a speech about wolf depredation as an impediment to the Western course of empire, Theodore Roosevelt placed his hand on the Bible and called the wolf "a beast of waste and desolation." The ensuing fusillade of government-sponsored predator control wiped out wolves in the contiguous United States, with the exception of northern Minnesota.[18]

Changing times brought changing conservation attitudes, informed by the Muries' and Leopold's science. In 1974, under the newly minted Endangered Species Act (ESA), USFWS put the gray wolf (*C. lupus*) on the federal list of threatened and endangered plants and animals, which is called *listing*. In 1976, USFWS listed the Mexican wolf (*C. l. baileyi*) as an endangered subspecies.

In chapter 3, I used the gray wolf as an example of how the ESA works to protect species and how species recovery takes place. As a result of the ESA, the federal government initiated wolf-recovery programs in two places: the Northern Rocky Mountain Recovery Area (NRM), which included Idaho, Montana, Wyoming, and the eastern third of Washington and Oregon, and the Blue Range Wolf Recovery Area (BRWRA), which comprised east-central Arizona and west-central New Mexico. By the 1990s, with approved recovery plans and the nonessential experimental rule, Section 10(j), which allowed wolves that threaten humans and livestock to be removed, the stage was set for the Yellowstone and Idaho wolf reintroduction.[19] However, even as wildlife agencies were preparing to *translocate* (i.e., capture and release) wolves to Yellowstone and central Idaho, gray wolves from Canada quietly dispersed south on their own and began to recolonize northwest Montana.

In 1995 and 1996, USFWS captured 66 wolves near Hinton, Alberta, and Fort St. John, British Columbia, and released 31 of them in Yellowstone National Park and 35 in central Idaho. Because the wolf subspecies that had originally lived in the NRM, *C. lupus irremotus*, had been extinct since 1926, the reintroduced wolves were *C. lupus occidentalis*—a different subspecies, but the same species, *C. lupus.* What it took to prepare to release those wolves fills 70 storage boxes in the USFWS archives at the National Conservation Training Center in West Virginia. I visited this wooded enclave on the Potomac River to examine these documents, guided in my work by USFWS historian Mark Madison.[20]

So what was the preparation needed to reintroduce wolves in the NRM? The short answer is hundreds of meetings by scientists and stakeholders and 160,000 public comments (100,000 pro; 60,000 con). These comments included hand-drawn wolf pictures from school children as well as angry letters from ranchers about how the government was reintroducing the wrong wolf species, all carefully catalogued and archived. The boxes contained scientific reports, decades of meeting notes, and many drafts of the evolving recovery plan. And in these boxes I also found the soul-stirring images of some of the first wolves released in Yellowstone and Idaho. The boxes told the amazing story of the NRM wolf reintroduction, a nearly 25-year effort, and they revealed that it simply wouldn't have happened without the steadfast effort of so many in the Service.

The NRM wolf reintroduction has been an unprecedented success. These wolves proceeded to do what came naturally to them: kill elk and reproduce. Although they sometimes killed livestock as well, such incidents were rare (less than 0.6 percent of livestock losses from all causes during the first decade). The wolf population grew quickly, by 2002 meeting recovery goals of at least 300 wolves and 30 breeding pairs in the three NRM states for at least three consecutive years. USFWS Director Dan Ashe called it "one of the great success stories of the Endangered Species Act."[21] Figure 5.1 depicts gray wolf distribution in 2013 within the Carnivore Way.

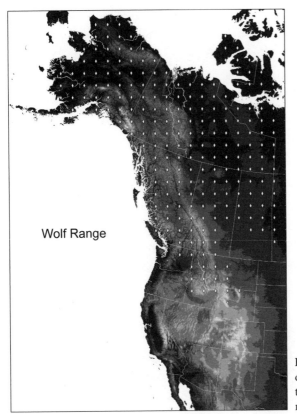

Figure 5.1. Map of wolf distribution in 2013 within the Carnivore Way. (GIS map by Curtis Edson.)

Wolf recovery has followed a very different trajectory in the Southwest United States. In 1977, USFWS contracted Roy McBride to capture wild Mexican wolves (*C. l. baileyi*) in Mexico to start a breeding program for this nearly extinct subspecies. He caught five animals over the next three years. By the early 1980s, humans had completely extirpated Mexican wolves (also called "lobos") from the wild. In 1990, USFWS hired David Parsons to lead the Mexican Gray Wolf Recovery Program. By the early 1990s, a panel of geneticists and taxonomists had certified other captive animals from the Ghost Ranch in New Mexico and the Aragon Zoo in Mexico City as pure Mexican wolves. Along with the wolves McBride had captured, this gave Parsons a total of seven wolves for the breeding program.[22] USFWS identified the Blue Range Wolf Recovery Area (BRWRA), which comprises about 7,000 square miles of public national forest lands in Arizona and New Mexico, as the best place for them. It contains a primary recovery area at its core, within a broader secondary recovery area.

A 1990 lawsuit by the Center for Biological Diversity led to the Mexican wolf's eventual reintroduction. In 1998, USFWS released the first eleven captive-bred Mexican wolves into the BRWRA. Since then, an irregular influx of captive-bred wolves has augmented this population periodically. Over the years, ranchers have actively opposed release of additional captive-bred wolves to buttress the wolf population. This opposition has made the process of Mexican wolf recovery more challenging. USFWS selects animals for release based on genetic diversity. Today, 50 licensed facilities breed wolves for this program. Leading facilities include the California Wolf Center, the Wolf Conservation Center in New York, and Ted Turner's Ladder Ranch in New Mexico. These centers also offer public education about wolves. Figure 5.2 depicts a captive Mexican gray wolf at the California Wolf Center.

Figure 5.2. Mexican wolf (*Canis lupus baileyii*) at California Wolf Center, Julian, California. (Photo by Cristina Eisenberg.)

In 2012, fourteen years after the first wolves were released in the BRWRA, their population had only reached 75. These wolves have struggled to survive due to higher human-caused mortality (e.g., poaching by ranchers, legal killing via the 10(j) rule) and a much smaller average litter size (two pups) than in the NRM. Experts such as Paul Paquet have concluded that, without key management reforms, this wolf population faces a very uncertain future.[23]

To learn more about lobo recovery, in October 2012, I spent a crisp autumn day afield with educator and filmmaker Elke Duerr. In 2011, she had made the film, *Stories of the Wolves: The Lobo Returns,* to help advance wolf conservation. She and I planned to look for wolf activity in the New Mexican portion of the BRWRA.

CʒƐɔ

We travel west on Interstate 40 from Albuquerque, and then south through the El Malpais National Monument (the name means "the badlands" in Spanish), and into Catron County, New Mexico. This 7,000-square-mile, economically-strapped county provides a home to 3,500 people, 12,000 elk, lots of cattle, a smattering of wolves, and some of the most virulent wolf hatred in the West. The Gila National Forest makes up a large chunk of Catron County. Wolves gravitate to the beautiful mixed open meadows and woodlands there, north of the Gila Wilderness. We decide to go into those woodlands to track wolves to see how they're faring.

We enter the forest on foot, following a narrow stream. Cattle sign far outnumbers other mammal sign. The cattle have trampled the riparian vegetation, causing the barren stream banks to erode. We spot coyote, elk, and deer tracks in addition to those of cattle. In a clearing next to the stream, we find a weathered, wolf-killed elk carcass. Its femurs have been bitten in two, the marrow sucked out—evidence of wolves, because in this area only they have carnassial teeth big enough to bite an elk femur in half. This carcass gives us hope, but after a thorough look, we don't find any other wolf sign.

We get back on the highway and continue south to Apache Creek, New Mexico, where we stop for gas. This pit stop provides big lessons about why Mexican wolf recovery is so challenging. Across the street from the gas station stands a billboard emblazoned with a lurid photograph of a dead calf and another of a dead horse. Both have had their bellies torn open. The billboard caption reads, "Beware, Wolves Nearby! Keep Kids & Pets Close!" This community has even installed cages for kids, to keep them safe from wolves as they wait for the school bus. I find this entrenched, irrational wolf-hatred sobering. It's irrational because south of the US-Canada border, wolf depredation accounts for a minor amount of total livestock losses, and no wild wolf has ever killed a human. It's sobering because these irrational fears are rooted in something very difficult to change: people's worldviews.

In Catron County, where the average income is $14,000 per year, 25 percent of the residents live in poverty, and 80 percent are ranchers, people despise the federal government. They've turned the wolf into a scapegoat for what they fear most: government control over what they can and can't do on their land. They fear being forced off public land, where some families have held grazing leases for several generations. The Apache Creek billboard depicts the powerful clash between the old and the new West. Unlike in the old West, where the only good varmint was a dead one, a growing number of living wolves populates the new West. People in places like Catron County are desperately trying to hold onto the mythological West and all its lawlessness, unfettered human freedom, and shoot-shovel-and-shut-up ways of dealing with wolves.

Old ways die hard. Defenders of Wildlife has made significant inroads in the BRWRA with programs such as their Wolf Compensation Trust, in which they reimburse ranchers for the value of cattle lost to depredation by wolves, and their Coexistence Program, in which they work to prevent depredation. To reduce conflict between wolves and people, they've helped ranchers purchase electric fencing, which stops wolves from entering livestock corrals. Other wolf deterrents include range riders, who ride among the cattle, and noise-making devices, called *rag boxes.* Some ranchers have accepted this help—and some have refused. When I ask Duerr what she thinks it will take to increase social acceptance of wolves here, she says, "Educating the next generation is our best hope."

Are Wolves Dangerous to Humans?

As the Apache Creek, New Mexico, billboard demonstrates, human risk keeps coming up in wolf-recovery arguments. Today's science is demonstrating that the real wolf is far less dangerous than the mythical wolf. Wildlife biologist Mark McNay found only nineteen cases of unprovoked wolf aggression toward humans between 1900 and 2000. The common factor in these incidents was that these wolves had become increasingly bold around people. Prior to documented attacks, wolves were observed stealing articles of human clothing and gear, exploring campsites, and obtaining food items.[24]

Scientists who have spent considerable time working with wild wolves haven't experienced aggressive wolf behavior. Doug Smith says,

> By a mile, wolves are the least dangerous carnivore. In order for them to attack humans, they almost always need exposure to people. Typically that's because they get fed and lose their natural fear of people, leading to an attack. This isn't the case with bears or cougars who may attack the first person they've ever seen in their lives. In other words, these other carnivores don't need to overcome the instinctive caution toward people that wolves have.[25]

Coexisting with Wolves without Fear: The Rewilding

Wolf recovery and delisting in the northern Rockies (NRM) provide valuable lessons about how we apply environmental laws such as the ESA. A key lesson has to do with how we define recovery. ESA recovery plans define it based on numbers: how many wolves or grizzly bears are needed to ensure survival of a population of a species for 100 years. However, as we've seen, recovery of a species such as the wolf has socio-political implications that transcend science. This can make the recovery road rather bumpy. In 2010, I co-taught a graduate policy class at Oregon State University with policy expert Norm Johnson, focusing on the wolf. During the course's ten-week span, NRM wolves were delisted and relisted twice, creating many teachable moments.

As we saw in chapter 3, by 2002, Idaho, Montana, and Wyoming collectively had met USFWS recovery goals of 30 or more wolf breeding pairs. Prior to delisting, the federal government required these states to create wolf-management plans. Idaho and Montana produced sound plans for meeting federal recovery goals (though these were considered too low by many experts). These plans called for at least fifteen breeding pairs in each state. Wyoming produced an unacceptable plan that called for dual classification of the gray wolf as a trophy game animal in wilderness areas and as a predator not subject to management or protection elsewhere. USFWS sent Wyoming back to the drawing board.

USFWS issued the first proposal to remove NRM wolves from the list of threatened and endangered plants and animals in 2007. The delisting proposal established a distinct population segment (DPS), which included Idaho, Montana, and Wyoming, plus eastern Washington and Oregon. Delisting called for removing gray wolf protection in the entire NRM DPS. In the ensuing lawsuit, a federal court found that USFWS had acted arbitrarily by approving Wyoming's wolf-management plan. In April 2009, when USFWS attempted to delist the NRM DPS, excluding Wyoming this time, the court found that the ESA didn't allow piecemeal delisting of a DPS. However, two years later, a rider to the 2011 Department of the Interior appropriations bill nullified the court's ruling and reinstated the 2009 delisting. The nullification meant the entire DPS was delisted. Since 2011, wolves have been hunted annually in Idaho, Montana, and Wyoming.[26]

Mexican wolf recovery has been far more difficult. BRWRA wolf numbers haven't approached the initial goal of 100. Parsons and others blame this in part on Section 10(j), which allows lethal wolf removal and has been over-liberally applied. The program has relied on releases of captive-bred wolves to buttress this population's limited genetic diversity and numbers. According to Parsons and John Horning of WildEarth Guardians, another issue has to do with the US Forest Service's unwillingness to retire grazing allotments within the BRWRA, despite willing ranchers who have found conservation buyers for their leases. Such easements would create safe havens for wolves. Many people see the battle over the lobo as being fundamentally about the struggle for control of Western public lands.[27]

In December 2010, USFWS appointed a new team to revise the Mexican Wolf Recovery Plan. The revised plan should incorporate best science, as mandated by the ESA. At the close of 2012, USFWS reported 75 wild Mexican wolves—the highest in years. However, by autumn 2013, a series of wolf deaths had demonstrated this population's precarious conservation status.[28]

When I ask Parsons what's needed for the lobo to recover, he says,

> Adhering to a science-based recovery objective, that's the short answer, I guess. And the other side of that is to establish rules that work for wolves. No doubt everything is going to be done under the "nonessential experimental rule," Section 10(j). I think it should be changed to "experimental essential" as a starting point. That would give the Service a little more clout to stick to their guns to make the numbers. It would make for a rule that's less flexible than the one we have, which has proven too flexible in the wrong hands.[29]

Experts believe that "best science," an ESA requirement, isn't being applied to the Mexican wolf, and that this failure may be hindering wolf recovery. Moreover, some question how in practice government agencies are defining "best science." Amaroq Weiss, who's worked on wolf recovery for decades says,

> Whose science is best science? In terms of best science, we have an insufficient amount of—and in some cases *no* amount of—peer-reviewed science that supports what most of us believe to be true: that you don't need to kill wolves to save them; that killing wolves and paying compensation doesn't increase human tolerance for wolves. In creating wolf-management plans, we field all of these topics, and yet there's very little peer-reviewed literature on them, and there needs to be more.[30]

In May 2013, the federal government filed a proposal to delist the gray wolf nationwide and maintain protection for the Mexican wolf by giving it both its own status as an endangered subspecies where found and also an expanded recovery area. Problems with this proposal include how one defines recovery of the gray wolf in "a significant portion of its range," as stipulated by the ESA. Experts like wolf biologist John Vucetich suggest that the NRM doesn't represent a significant enough portion of the range of a species that once covered most of North America to allow us to call that species recovered. The delisting proposal leaves newer wolf populations in Oregon and Washington with no federal protection, thereby jeopardizing their long-term survival. Regarding the Mexican gray wolf, while giving this subspecies its own endangered status and expanding its recovery area will help, this proposal doesn't address its high mortality, conflicts with humans, and low reproductive rate, nor does it acknowledge the best science from USFWS's own recovery team on how many wolves should be recovered and where.[31]

Wolves and Hunting by Humans: Lessons from the Far North to Yellowstone

Wolves have never gone extinct in the far north. Designated a trophy or fur-bearing species outside national parks, the wolf is subject to regulated, widespread killing by humans, as well as predator control. Wildlife-management agencies and state and federal governments use the term *liberal take* for this. In this context, *liberal* means that there are few restrictions, and *take* is a wildlife-management word for killing that has been codified into legal language (for example, *take* appears in the ESA and other environmental laws).[32] Wildlife-management agencies use the term *intensive management* to refer to keeping wildlife numbers, primarily carnivore numbers, as low as possible. This is achieved mainly by aerial shooting. Subsistence use is intrinsic to human survival in the far north, an extreme landscape climatically. However, because plants grow very slowly in the far north, it's difficult to provide enough food for ungulates to sustain the high populations that the hunting public demands. To boost ungulate numbers, the Alaska Department of Fish and Game keeps wolf populations stable overall but low regionally, and the agency even eliminates them completely in some places. Since the 1980s, Alaska has had approximately 7,000–11,000 wolves, so extinction isn't an issue. Wolves are managed similarly in the Yukon.[33]

Not everyone agrees that intensive management works. Bob Hayes spent eighteen years as the wolf biologist for the Yukon Fish and Wildlife Branch, killing over 800 wolves during his career. Initially, he approached wolf control as an ecological experiment to help make better management decisions. He found that after an initial sharp increase in moose and caribou, as soon as wolves recovered, their prey began to decline again, necessitating another round of intensive management. Or, with fewer wolves, moose exploded in number, depleted their food supply, and then died en masse from starvation and disease. Additionally, he and others found that ungulate numbers can have as much to do with climate as with predation. Hayes says, "Science clearly shows killing wolves is biologically wrong. When we kill wolves, we're killing the very thing that makes the natural world wild."[34] John Vucetich and ethicist Michael Nelson agree. Further, they suggest that the ethical implications of gunning down a living being from an aircraft are obvious if one recognizes that wolves have *intrinsic value* (i.e., value simply because they're sentient creatures).[35]

Wolf-management practices vary in western Canadian provinces. Well tolerated in southern Alberta, wolves receive nearly zero tolerance in the northern portion of the province. Here and in British Columbia, managers remove 60–80 percent of the wolf population annually by poisoning with strychnine and aerial shooting, in order to protect woodland caribou (*Rangifer tarandus caribou*), listed as "threatened" under Canada's Species At Risk Act (SARA). Habitat loss due to mining, logging, oil and gas exploration, and motorized recreation all contribute to caribou decline. Management solutions have focused on wolf eradication to protect caribou, rather than on limiting human activities that negatively affect caribou habitat. This has created a domino effect. Deer, which thrive in disturbed places, have increased due to the young plant communities that arise there. More deer means more wolves, which primarily feed on deer, but take caribou opportunistically.

෬෪

When it comes to wolves, few places dominate the headlines like Yellowstone National Park. This iconic park has been likened to a conservation theater, where wolves provide the drama. Since the 1995–96 wolf reintroduction, millions of people have visited the park to observe wolves. Here as elsewhere, we're finding that wolf recovery is as much about people as about this apex predator's ecology.

On any given day, if you drive through Lamar Valley you'll see dozens of vehicles at the pullouts along the park road. People with spotting scopes line up to get glimpses of wolves engaged in their daily lives: raising pups, hunting elk, fighting grizzly bears over carcasses. The media attention these wolves have drawn has brought their ecological role into the public eye. People who watch wolves contribute to conservation by also observing and bearing witness to how wolves are creating a healthier ecosystem here. As an added and important benefit, wolf watching contributes to the vitality of the human communities outside the park, through tourism dollars.

Rick McIntyre, a Yellowstone wolf biologist who since 1995 has spent over 5,000 days observing these wolves, facilitates wolf viewing and enormously contributes to Yellowstone wolf ethology studies. Aided by longtime volunteers Laurie Lyman, Kathie Lynch, and Doug McLaughlin, McIntyre works with Doug Smith, Dan Stahler, Yellowstone Wolf Project staff, and Jim Halfpenny to support research and raise public awareness of wolf natural history. Since their reintroduction, McIntyre has found that Yellowstone's wolves have been powerful teachers.

Norm Bishop, who as the park's former chief of interpretation led the public education aspect of the reintroduction, believes that, since bringing wolves back to the park, we've made a quantum leap in our knowledge of them. He says, "Before the reintroduction, Congress asked questions about the effects of wolves on game, on grizzly bears, on livestock. We had to go to Alberta and Minnesota for answers. We assembled a group of experts to answer questions about things like wolf sub-speciation and how wolves would impact wildlife diseases. And we could speculate about the potential effects of wolves in Yellowstone and model these effects from known cases, but we had to have wolves on the ground to find the answers."[36]

Yellowstone National Park has changed tremendously since wolves returned. Emmy Award–winning filmmaker Bob Landis has chronicled these changes. Films he shot in the late-1990s depict short aspens and shrubs suppressed by decades of heavy elk browsing. Before wolves, the park had a policy of natural regulation, which meant letting nature take its course. In the early 1990s, the Northern Range (an elk wintering ground that includes the Lamar

Valley) had nearly 20,000 elk and no wolves. With no wolves present, elk were browsing woody plants to death. But by the early 2000s, after the wolf's return, Landis's films, such as *In the Valley of Wolves*, show a different picture—aspens and shrubs beginning to grow taller than elk can reach. As we saw in chapter 2, this height increase had to do with big changes in elk behavior as a result of learning to live with wolves.

Yellowstone elk had a short, fast learning curve in response to wolves. Their lessons had to do with the fact that in Yellowstone as elsewhere, no other predator is capable of killing adult elk as often and efficiently as wolves (in 2012, elk made up 63 percent of Yellowstone wolf diet). To avoid being killed by wolves, elk had to learn a variety of strategies: making larger groups, making smaller groups, going into the forest for shelter. Those with the confidence to stand their ground when approached by wolves weren't attacked. Those that felt vulnerable ran, making them easier to kill. They learned to avoid places with deeper snow, where they'd be easier targets for wolves.[37] Over the years, Landis documented wolves refining their hunting strategies to keep up with savvier elk.

People learned big lessons, too. As wolves recovered, lost their ESA protection, and began to be hunted by humans outside the park, observers from Smith to park visitors learned that this apex predator's recovery isn't just a numbers game. Nevertheless, numbers are a good place to start. From 1995 on, wolf numbers grew rapidly. In the early 2000s, graphs of this wolf population looked like the letter "J," with growth reaching skyward in an optimistic curve. But the boom couldn't last forever. In nature's economy, resources, or carrying capacity, limit predators and their prey.

In Yellowstone National Park as elsewhere, wolf and elk fortunes are inextricably bound to each other and to human activity. In 2004, the park's wolf population peaked at 174 individuals and then began a downward trend. From 2007 to 2012, wolves declined by 50 percent to 83 individuals, due to disease, decline of the elk herd, and hunting by humans. Based on radio-collar data, Yellowstone wolves spent 90 percent of their time within the park. However, when they stepped outside the park, they could be legally hunted—and they were, starting in 2009, with a break in 2010, and continuing since 2011. At the close of the 2012/13 wolf-hunting season, Yellowstone's wolves had dropped to 70 individuals—a 15 percent decline since 2011. Meanwhile, from 2007 to 2012, the elk population declined by 42 percent, to 4,000 individuals. The park attributed elk decline to drought, predation by wolves, predation by grizzly bears on elk calves, and to the late-winter cow elk hunt outside the park, which resulted in the death of pregnant females.[38]

The impacts of the wolf hunt created a debate about how hunting wolves outside the park is affecting their ability to thrive inside the park. In the NRM after USFWS removed ESA protections from wolves in 2009, Montana and Idaho set annual wolf-hunt quotas of 20 percent of the regional wolf population. In 2009 in the two states combined, humans killed 37 percent of the wolf population through legal and illegal hunting as well as management actions (e.g., for livestock depredation).[39]

Understanding the effects of the wolf hunt requires awareness of basic population ecology. *Offtake* (also called *harvest* or *take* by wildlife agencies) refers to the total number of individuals removed from a population via hunting or management actions. Two key concepts frame the argument about wolf offtake: *compensatory* and *additive* mortality. If wolf offtake is compensatory to natural causes of death (e.g., disease, age), then a natural reduction in other types of mortality compensates for offtake. This means that hunting wolves won't cause the wolf population to decline further than it normally would if we didn't hunt wolves. Conversely, if wolf offtake is additive to natural causes of death, hunting wolves will cause the wolf population to decline more than it would if we didn't hunt wolves. Furthermore, as a population falls below carrying capacity, offtake becomes more additive and less compensatory. In other words, the lower a wolf population is to begin with, the greater the effect hunting will have on the population.

Some scientists see offtake as primarily compensatory. After decades of studying wolves in the Upper Midwest and Alaska, Mech believes that it takes an annual offtake of over 50 percent to suppress a wolf population. In Alaska's Brooks Range, because of wolves' high reproductive potential and resilience, wildlife biologists such as Layne Adams recommend continuing a modest wolf offtake.[40]

Other scientists see offtake as primarily additive. In the NRM, Scott Creel and Jay Rotella conducted a meta-analysis of the relationships among human hunting of wolves, total wolf mortality, and population dynamics. They found that while wolves can be harvested sustainably within limits, killing wolves by hunting is generally additive, because when we kill older wolves, younger animals get into more trouble—the way that, in a human family, if mom and dad suddenly died, teenagers would get into more trouble. Dennis Murray and colleagues found additive wolf mortality in the NRM due to hunting, especially where conflict with livestock occurred, but found partial compensatory effects as the wolf population increased. They also found that offtake may have a more additive effect on juvenile and dispersing wolves.[41]

Collectively, these studies indicate the need for caution when developing policies for wolf hunting and trapping. The lack of consensus in these studies, which are being applied to the population of a species that until recently had ESA protection, means that perhaps we don't quite yet know what "best science" is in this case, and so killing wolves without knowing the full ecological consequences of reducing their numbers may be unwise.

But the impacts of the wolf hunt aren't just about numbers. While a 12 percent drop in the Yellowstone National Park wolf population from 2012 to 2013 due to the wolf hunt and other causes of mortality may be biologically sustainable, it may have other negative consequences. According to Smith, Yellowstone's wolves used to be one of the few unexploited wolf populations in North America. Before the wolf hunt, it experienced an annual human-caused mortality (e.g., from vehicle strikes) of less than 4 percent. Because hunting these wolves outside the park affects their behavior inside the park, the wolf hunt has ended the invaluable opportunity to study an unexploited wolf population as a scientific control against which exploited populations could be compared.[42]

By hunting wolves we disrupt their society and destabilize their packs. Packs may split into smaller packs made up of younger animals, with a greater influx of unrelated individuals. And younger, less-complex packs may kill cattle or approach humans for food.[43] The Lamar Canyon pack, formerly one of the most stable and viewable park packs, provides a case in point. When the gun smoke cleared from the 2012/13 wolf hunt, this pack's story provides a cautionary tale about the unintended consequences of hunting wolves immediately outside national parks.

Before the 2012/13 hunt, an illustrious pair led the Lamar Canyon pack: wolf 832F, whom biologists called the "'06 female" (for her birth year) and wolf 755M. But pack leadership also included wolf 754M, the beta male (and 755M's brother), who'd vied for 832F's admiration and then helped the alpha pair care for their pups. This trio engendered tremendous public affection. Capable of taking down an elk by herself, '06 quickly became a legend. She ranged widely through the Lamar Valley, yet she seldom left the park.

Tragedy struck in November 2012, when 754M met a hunter's bullet in Wyoming, outside the park. The next month, the '06 female also went down in the wolf hunt. Both were big wolves, the kind that trophy hunters like to get in their crosshairs. Their entirely legal deaths played out publicly and created public outrage. But that was only the beginning of the trouble.

When breeding season began in late December, one of '06's daughters, the second-ranking female in the pack, became the new alpha. However, because she was the alpha male's daughter, she wouldn't breed with him. And so now 755M, the alpha male, a great hunter with the most social experience, went looking for a mate. This left the Lamar Canyon pack unstable and leaderless going into breeding season.

By late January, the Lamar Canyon pack was going through major changes. Two of 755M's daughters had attracted mates from other packs. Meanwhile, 755M, who'd been wandering, had found a mate, 759F from Mollie's pack, and returned to his pack with her. But pack dynamics had shifted in his absence, so what he returned to was actually partly his old pack with some new wolves. The new males turned on 759F, killed her, and ran 755M off. By April 2013, 755M's daughters were both pregnant and preparing for birth, and the pack was spending a substantial amount of time outside the park, in Wyoming. Biologists occasionally observed 755M in Lamar Valley with two other wolves.[44]

Eventually '06's daughters had their pups. While at first this seemed an example of wolf resilience, in that these wolves reproduced and continued to maintain a pack, further events demonstrate how '06's and 754's deaths disrupted this pack's social stability. For example, in August 2013, wolf 820F, '06's two-year-old daughter, left the pack under hostile pressure from her older sisters. That she'd spent her entire life in the park and was very used to people led her to make a foolish choice. She started spending a lot of time in Jardine, Montana. When she turned to raiding chicken coops for food, Montana Fish, Wildlife, and Parks (FWP) killed her. While the Lamar Canyon pack's ultimate fate remains uncertain, its short-term response to the wolf hunt demonstrates both resilience and instability in the face of challenges.

These deaths and others occurred because, for the 2012/13 hunting season, FWP removed the partial buffer outside the park where wolves couldn't be hunted. This meant that, for the first time, wolves could be hunted immediately outside the park with no quotas. After the deaths of the '06 female and 754M in late 2012, the FWP Commission tried to close two areas adjacent to the park to hunting and trapping because too many Yellowstone wolves were being killed. Anti-wolf groups sued, accusing FWP of giving them insufficient time to comment. A Montana judge issued an injunction, and FWP removed the buffer, leaving park wolves vulnerable in places like Gardiner, Montana, an elk wintering ground immediately outside Yellowstone.

Lifting the buffer created an ethics problem. Lamar Valley wolf-viewing has made these wolves tolerant of and even habituated to humans. Much wolf-viewing takes place within 200 yards of a wolf—the range of a rifle shot. This makes hunting park wolves outside the park not entirely fair to the animal being hunted. Similar issues have arisen in Denali National Park. Coinciding with the state's decision to remove the buffer zone, wolf numbers have dropped to their lowest level in decades.

In Denali, park biologist and University of Alaska Fairbanks PhD student Bridget Borg, (who was with us during our visit to the East Fork den in that park) is studying the effects of the wolf harvest (i.e., trapping) and buffer on visitor sightings of wolves within the park. While wolf harvest outside park boundaries may not have negative long-term effects on the population of a species as resilient as the wolf, its impact on local packs that provide the majority of park sightings may be significant.

Borg's work involves a network of people and incorporates citizen science. Staff who drive the buses along the Park Road, as well as trained NPS personnel and volunteers, report wolf and other wildlife sightings. While the park has been documenting such sightings for years, more recently this system has been digitized via data-logging units inside the buses and handheld data-loggers for other observers. Borg uses sighting data along with wolf GPS-collar and remote-camera data to identify places where wolves may be more viewable.

She's finding that local wolf population size, breeder loss, and proximity of a wolf den to the road best explain wolf viewability. For example, loss of breeding wolves due to trapping outside the park can reduce the reproductive success of park packs. This in turn lowers the viewability of wolves inside the park. Since the buffer was removed, wolf viewing has plummeted. The wolf-sighting index (a measure of how wolf sightings change from year to year) has dropped from 44 percent in 2010, to 21 percent in 2011, 12 percent in 2012, and only 4 percent in 2013. In the park and preserve north of the Alaska Range, spring wolf counts went from 66 in 2012 to 55 in 2013—the lowest since counts began in 1986.[45]

Millions of people visit Denali and Yellowstone each year. Most come to see wolves. The drop in viewability caused by wolf hunting and trapping is negatively affecting tourism—the lifeblood of the small towns near these iconic parks. The link between the buffer zone and tourism has the ability to make or break local economies. Thus, implementing harvest buffer zones around these parks creates a win-win situation in which wolves *and* people benefit.

In a 2014 *Washington Post* photo essay entitled "The Last Wolves?," noted wildlife-behavior expert, conservationist, and humanitarian Jane Goodall comments on the need to protect park wolves in both Denali and Yellowstone from hunting and trapping outside park boundaries. In the text that accompanies Thomas D. Mangelson's photographs of wolves in Denali and Yellowstone National Parks, she says,

> Those responsible for managing wolf populations typically think in terms of the species as a whole; they calculate the number of wolves that an environment can support or lose. But it is important to remember that each pack is composed of bonded individuals. When leaders are killed, the pack and its traditions may disintegrate. When breeders are killed, the pack's survival is threatened even more directly.[46]

Recovery of the Wolf

We stand at a wolf crossroads. As of 2013, more than 50,000 wolves existed in North America. The recent USFWS proposal to remove ESA protection from the gray wolf throughout the conterminous United States, exclusive of the Mexican wolf and the red wolf (*Canis rufus*), gives us a choice. We can choose to coexist with wolves, or we can reduce their numbers to a point where their ecological benefits may be negligible and they hover just above extinction. This choice is made more difficult for some people because the costs of living with wolves (e.g., a wolf eating a rancher's cow) are easy to quantify, while their ecological benefits (e.g., young aspens growing into the forest canopy) are not.

Sharing our world peacefully with wolves has less to do with the animal and more to do with humans. As a young Parks Canada biologist in the 1970s, Kevin Van Tighem watched wolves return. Despite this recovery, he remains concerned. "Wolves live in a physical landscape affected by human decisions, and they also live in a human social landscape. And that social context is really troubling. I don't know how we can create the kind of society that can sustain its carnivores. They're our competitors so it's complicated enough to begin with, without lack of civil discourse between humans."[47]

In order for wolf recovery to progress, we must create a practical and respect-based conservation strategy. In western North America, conflicts with wolves occur on the southern hem of their range. And it's in these places, where twenty years ago no wolves existed, and where today people have to learn to live with them, that conservation organizations have focused their work. Dozens of nonprofit organizations (NGOs) and thousands of people are working on wolf conservation. Here are some examples of inspiring grassroots efforts:

Carter Niemeyer has spent his lifetime in the trenches with large carnivores. Literally. A lifelong trapper, hunter, and wildlife conservationist, for 26 years he worked for the US Department of Agriculture Damage Control, where he was a *wolfer* (one who kills wolves). He retired in 2006 from USFWS as the Idaho federal wolf-recovery coordinator. During the Northern Rockies wolf reintroduction, he emerged as a leading wolf proponent. Since his retirement from government work, he's been an outspoken critic of our wildlife-management system. Today he works with Defenders of Wildlife and Living with Wolves to encourage more-accurate accounts of wolf depredation and to improve the use of nonlethal depredation deterrents (e.g., range-riding, electric fencing, and *fladry*—rope lines carrying small flags that are highly visible and audible to wolves). With regard to the threat wolves present to people, Niemeyer says, "Wolves do not, as many believe, kill everything in sight, destroy their own food supply, or lick their chops at kids waiting at bus stops. They are simply predators like lions and bears, and anyone who believes otherwise is, well, wrong."[48]

Timm Kaminski is a founder of and principal investigator of the Mountain Livestock Cooperative, which works to sustain working ranches and large carnivores. He and Alberta rancher Joe Englehart found that managing cattle to mimic wild ungulate behavior, such as keeping the herd together to avoid single animals being targeted, reduces and even eliminates depredation. He recommends that ranchers spend more time riding with their cattle, keeping them more tightly bunched, and moving them across rangelands, which also helps improve range conditions. Since 2012 Kaminski has partnered with wildlife biologist Kyran Kunkel and Living with Wolves to help advance coexistence in rural communities.

Louise Liebenberg owns and operates Grazerie, the first certified predator-friendly sheep ranch in Alberta. She says being predator-friendly "is about responsibility, biodiversity, and simply about sharing the land with the wildlife that inhabits this country." To keep her 600 sheep safe, she breeds Sarplaninac shepherds, a strong-willed Yugoslavian dog breed that instinctively protects livestock. Grazerie maintains eight to ten Sarpalaninacs. When a predator comes calling, the dogs form a circle around the sheep, facing outward. Older, more experienced dogs advance, while younger dogs stay with the sheep. Thanks to her sheepdogs, Liebenberg hasn't experienced any wolf depredation.[49]

Steve Clevidence's family homesteaded Montana's Bitterroot Valley in the mid-1880s. Ranching dominates the local economy of this strongly anti-wolf region. Clevidence, a fifth-generation rancher, is one of many in his family who've prided themselves on good stewardship that includes Native American practices (called Traditional Ecological Knowledge). He keeps cattle together to avoid single animals being targeted by wolves and feeds them in the evening, so they bed down overnight. He also listens for wolf howls and uses traditional wildlife tracking to look for them. If he detects wolves, he moves the cattle to other parts of the ranch. More recently, he's been mentoring Dan Kerslake, a young predator-friendly Bitterroot Valley rancher, to help spread his knowledge to new generations.[50]

Some ranchers go into academia to gain a better understanding of coexistence. Canadian rancher Olivier LaRoque recently completed a PhD in anthropology, studying how we manage private and public lands and the role of NGOs in improving ranching practices. He's found that coexistence with wolves is mostly about politics. He says,

> The outstanding question about wolf conservation is whether it is possible to share a landscape with wolves without being continuously at war with them, especially when there is livestock in the mix. There is much speculation all around, that it is possible to get along with wolves, or that wolves are incurable fiends, or that they are necessary to restore the "balance of nature." The jury is still out on every account. The outcome of an interspecies social contract really depends on the fine print.[51]

CR80

Few creatures so resemble humans and our societies as does the wolf. Perhaps we can find our way to a truce with this apex predator that can teach us how to live more ethically and sustainably on the earth. We've made good inroads, yet still have a long way to go. However, as insurmountable as anti-wolf intolerance may seem, at least we have some control over it. In the next chapter we'll learn about another large carnivore species whose welfare is primarily threatened by something we have little control over: climate change.

Wolverine (*Gulo gulo*). (Drawing by Lael Gray.)

CHAPTER SIX

Wolverine (*Gulo gulo luscus*)

Not long ago, a group of friends in a national park in the Canadian Rocky Mountains took a hike on a popular trail known for high grizzly bear (*Ursus arctos*) activity. A couple of miles along the trail, they had an encounter with what appeared to be a grizzly. The animal came at them growling and hissing, making horrific sounds. All in the group climbed trees to get away from it, except for one person, who was able to flee to the warden office for help. Upon arriving at the incident site, a park warden discovered that the animal involved was actually a wolverine (*Gulo gulo luscus*). The astonished hikers reported that it had behaved so ferociously that they'd been certain it was a grizzly. While wolverines don't normally threaten humans, this story illustrates the species' formidable nature.

This incident wasn't the first time people have mistaken these chimeric creatures for bears. Over the centuries, wolverines have acquired a reputation for ferocity and strength well beyond their size. In North America, going as far back as the early 1800s through the present time, records abound of them driving grizzly bears from carcasses, fighting off wolf (*Canis lupus*) packs, and killing thousand-pound moose (*Alces alces*). According to wildlife biologist and

renowned journalist Doug Chadwick, who volunteered on the Glacier National
Park Wolverine Research Project for several years,

> If wolverines have a strategy, it's this: Go hard, and high, and steep, and never back
> down, not even from the biggest grizzly, and least of all from a mountain. Climb
> everything: trees, cliffs, avalanche chutes, summits. Eat everybody: alive, dead,
> long-dead, moose, mouse, fox, frog, its still-warm heart or frozen bones. Whatever
> wolverines do, they do undaunted. They live life as fiercely and relentlessly as it has
> ever been lived.[1]

Wolverines are the stuff of legends. People who spend the most time with
them—scientists—are purveyors of some of the best tales about them. Scientists describe wolverines as curious, canny, tireless, and indomitable. When
trapped in the specially designed log boxes used to hold them so they can be
radio-tagged, the sounds they make resemble a cross between a chainsaw and
a Harley Davidson motorcycle. If not pulled out of box traps in a matter of
hours, they'll claw and chew their way out. In their daily lives, wolverines per-
form physical feats the likes of which few creatures are capable, and with what
appears to be effortless grace. For example, in Glacier National Park, a young,
radio-tagged male wolverine summited Mount Cleveland—the park's highest
peak at 10,466 feet—ascending the last 4,900 feet up a sheer, nearly vertical ice
rampart in less than 90 minutes. He made the ascent for no obvious reason and
presumably left his urine on the summit to mark his turf. That this animal's extreme walkabout was documented empirically via a radio collar—which com-
municated with several global positioning satellites (GPS) in space, which in
turn transmitted data to a researcher's laptop, where software turned the location coordinates into dots on a raster-celled topographical map—only increases
our amazement at such acts. Even the most staid scientists refer to the wolverine
in the vernacular as a "badass" species. And if you carefully read scientific journal articles about wolverines, you find embedded in them a healthy measure of
respect for the species.[2]

Despite all this fearlessness and force, the wolverine is one of the most at-
risk species south of the US-Canada border. Factors such as climate change,

trapping, and human disturbance represent the biggest threats to its conservation. Given our shrinking wilderness and warming planet, the wolverine's plight emphasizes the need to conserve corridors along the Carnivore Way. Maintaining a web of safe pathways and habitat linkages amid current policies of land use and human development is critical to conserve this species. As with the other large carnivores whose ranges include the far north, wolverine persistence is most threatened in the southern portion of its range, in Washington, Idaho, Montana, Wyoming, and Colorado.

What's in a Name: Double Glutton

Names have a strong ability to shape human attitudes about anything. All taxonomists label things. Early taxonomists took liberties with these labels, often giving organisms names filled with value-charged meanings. (Today, such practices are unusual.) As soon as you label something, you change how it's perceived. The wolverine's taxonomy provides the first clues about some of the conservation challenges it has faced throughout the centuries.

The wolverine, *Gulo gulo*, is the largest terrestrial member of the weasel family (Mustelidae). In 1780, German taxonomist Peter Pallas gave this king of the mustelids its scientific name, which means "double glutton," a name laden with baggage. The Roman Catholic Church considered gluttony one of the Seven Deadly Sins. Unsurprisingly, the link between gluttony and sin led to widespread wolverine persecution.[3]

A holarctic species, the wolverine originated in Eurasia and crossed the Bering Land Bridge 10,000 years ago. Two subspecies of wolverine exist: *G. g. gulo*, commonly known as the Old World wolverine, distributed in northwestern European countries, Siberia, and Russia; and *G. g. luscus*, the New World wolverine, distributed in North America.[4]

Native Americans have mixed perspectives on the wolverine. While some nations hold it in high regard, others revile it. For example, the Cree call it *ommeethatsees*, which means "one who likes to steal," in reference to its plunder of traplines and skill at ransacking cabins for food. The Algonquian Micmac call it *kwi'kwa'ju*, which means "malevolent spirit." French fur trappers mispronounced that as *carcajou*, which became a common alternative name for the wolverine.

European settlers gave this animal derogatory names such as "skunk bear," for its musky odor, and "mountain devil," for its ferocity. The wolverine's calling card—the severed feet of the animals caught in raided traps left reeking with musk so that everyone would know the perpetrator's identity—only reinforced these negative associations. For the first European settlers in the New World, survival was so close to the bone that they saw anything that competed with them for resources as a threat to be obliterated. And so it was for the wolverine.[5]

Wolverine Natural History

One day, while I measured aspens in an avalanche chute in Glacier National Park, Montana, a stocky animal as large as a medium-size dog streaked across my path. All I got was a fleeting look, enough to see the pale-tan chevron emblazoned on its chocolate-brown fur. Stunned, I realized it was a wolverine—considered little more than a phantom due to its scarcity. Indeed, in over two decades afield in the northern Rocky Mountains, this is the only one I've seen.

Wolverines are one of the rarest, least-known terrestrial mammals. They reside amid snow peaks and glaciers, where they cover vast amounts of territory. This makes them intractable for research for all but the most intrepid scientists. Radio collars slip off their short, thick necks. Track surveys from aircraft have proven effective, but expensive.[6] Given these challenges, as of 1994, only four wolverine studies had been completed in North America. However, since then, wolverine research has proliferated, with a focus on developing more-accurate population censusing methods. Advances include GPS collars (which, despite a tendency to slip off wolverines' necks, still provide the highest quality, fine-scale data on animal movements), VHF transmitter abdominal implants, non-invasive survey methods such as remote cameras, DNA analysis of snagged hair, and population modeling.[7] The resulting stack of peer-reviewed scientific journal articles are creating a greater understanding of this species.

Wolverines live in boreal, taiga, tundra, and alpine biomes. Low-temperature extremes and/or high-elevation extremes characterize their habitat. They've adapted to live in and exploit this unproductive niche that tests the limits of mammalian existence. To do so, they maintain low population densities and reproductive rates, large home ranges that they patrol relentlessly, and a social system that gives them unique access to foods in their territories.[8]

Wolverines are superbly built for their fierce lives. They have a broad, wedge-shaped head, blunt nose, and round ears. Their eyes are relatively small for the size of their head. Like all members of the weasel family, they have sturdy legs and a muscular body. Highly water and frost resistant, their thick fur can range in color from near black to cinnamon. Light brown lateral stripes along both sides of the body give their pelage a cape-like appearance. However, there's great variation among individuals, with some wolverines having more white than dark fur. Their plantigrade gait enables them to walk on the full soles of their feet, like a bear or a human. Crampon-like claws on their five-toed feet facilitate climbing, digging, and moving through deep snow. Pound for pound, they have a more powerful bite than most other carnivores—greater than the wolf's and similar to the Tasmanian devil's (*Sacrophilus harrisii*). For nourishment, wolverines can crunch up the bones they find as they roam their territory.

Wolverine habitat choice is driven mainly by snow availability. They live in places with snow cover most of the year. South of the Artic and subarctic, this means elevations between 7,200 and 8,500 feet. Below the US-Canada border, in summer, wolverines favor whitebark pine (*Pinus albicaulis*) habitat in the subalpine zone. As snow accumulates in winter, they spend time at lower elevations (sometimes as low as 4,500 feet), in Douglas fir (*Pseudotsuga menziesii*) and lodgepole pine (*Pinus contorta*) forests. They prefer north-facing slopes, where snow accumulates first in winter and lingers longest in spring. Jeff Copeland and his colleagues found that, worldwide, 90 percent of the time, radio-collared wolverines were found in places that had a maximum August temperature of less than 72 degrees Fahrenheit.[9]

With one of the lowest reproductive rates of any large carnivore species, wolverines have very specific requirements for successful reproduction. This polygamous species mates once per year. Both males and females become sexually active at the age of two years, and females can reproduce by the age of three. They breed but seldom whelp annually. Breeding occurs from summer through fall, with embryonic diapause, an evolutionary adaptation they share with grizzly bears, in which fertilized eggs can wait in a suspended state for months. If a healthy female has sufficient food, the eggs will implant on her uterine wall and begin to develop into wolverine kits. But because of the high energetic costs of reproduction, a malnourished female will reabsorb the embryos rather than give birth. Gestation ranges from 30 to 40 days, with birth occurring as early as January or as late as April. In Glacier National Park, considered prime wolverine habitat, Jeffrey Copeland and Richard Yates found a 68-percent pregnancy rate. Wolverines have small litters of one to four kits. However, as a female matures she may produce bigger litters, provided she has abundant food. Wolverines' low reproductive rate makes the species even less resilient than grizzly bears.[10]

Pregnant females need deep snow for denning. They dig two types of dens: *natal dens*, in which they give birth, and *maternal dens*, in which they raise their young. Sometimes excavated under fallen trees or boulders for structural support, both den types have multiple entrances, long, branching snow tunnels, and several beds. After giving birth in natal dens, females often move their unweaned kits to one or more maternal dens. During weaning, which occurs at around ten weeks of age, kits leave the den with their mother. This coincides with periods when the maximum daily temperature rises above freezing for several days in a row. Females sometimes use additional dens after weaning, called *rendezvous sites*, to keep kits safe from male wolverines from other territories. These intruders often establish genetic dominance by killing other males' offspring.[11]

The emerging picture of wolverine relationships with their own kind differs considerably from the long-standing image of this species as antisocial and highly aggressive. Recent studies are revealing a complex social structure in which parents often share responsibilities for their offspring—a form of wolverine "joint custody." Long after weaning, offspring sometimes spend time with their parents, learning to hunt and scavenge. In Glacier National Park, GPS data has shown wolverine offspring joining their father to learn to hunt, fathers visiting maternal den sites, even bringing food to the female, and adult siblings staying together for a while, then separating and reuniting periodically over multiple years. Researchers also have observed a male wolverine traveling in a group that consisted of yearling and adult offspring—three generations of this wolverine family.[12]

Wolverines cover a lot of ground. Their home ranges vary from 39 to 460 square miles. Two main factors influence home range size: food availability and gender. Home ranges are larger in poor environments that have less food and also larger for males than for females. In the far north, wolverines have smaller home ranges in areas with abundant predator-killed ungulate carcasses. In the Greater Yellowstone Area (GYA), which includes portions of Idaho, Montana, and Wyoming, Robert Inman, director of the Wildlife Conservation Society's (WCS) Greater Yellowstone Wolverine Program, found home ranges of 308 square miles for males and 117 square miles for females. Because neighboring wolverines tolerate females better than males, females often establish a territory next to their mother's.[13]

When ranging over their turf, wolverines scent-mark as they go, in effect outlining their holdings. This helps them maintain exclusive territories for breeding and feeding, maximizing access to food in a highly resource-limited system. Until recently, we thought wolverines marked their territory with musk, an oily, strong-smelling substance. However, researchers have discovered that they actually use urine for scent marking. Wolverine urine contains terpenes, pungent compounds associated with the pine needles that are part of their diet. They discharge musk involuntarily, more for defense than for marking.[14]

Moving with an inexhaustible lope, wolverines make prodigal movements and long-distance dispersals. M304, a young male in Inman's GYA project, traveled 550 miles in 42 days, moving south from Grand Teton National Park, east to Pocatello, Idaho, then to northern Yellowstone National Park, and back to the Tetons. And in April 2009, M56, another young male captured, radio-collared, and released by Inman, dispersed over 500 miles across the Great Divide Basin. He crossed Interstate 80 on Memorial Day weekend and turned up in Rocky Mountain National Park, Colorado.[15]

Opportunistic omnivores in summer and primarily scavengers in winter, wolverines subsist on a variety of foods. Their prey includes marmots (*Marmot caligata*), snowshoe hares (*Lepus americanus*), porcupines (*Erethizontidae* spp.), beavers (*Castor canadensis*), and lynx (*Lynx canadensis*). However, they also scavenge these species as carrion. Other key wolverine foods include small mammals, birds, pine needles, berries, and insects.[16]

Their prowess at finding and taking food earned wolverines their scientific name. Dead or alive, from fresh flesh to bare bones, wolverines will eat it all. Like all members of the weasel family, because of their high metabolic rate they must eat much to survive. Their large stomachs can hold up to twenty pounds. Gorging provides a hedge against food scarcity and competition from other wolverines. But despite their scavenging skill, wolverines have a precarious existence. In a meta-analysis of twelve wolverine radio-telemetry studies conducted between 1972 and 2001, John Krebs found starvation a leading cause of mortality. A carrion-eater's life isn't easy. While wolverines can chase grizzlies off carcasses, bears can and do kill them. Similarly, cougars (*Puma concolor*) sometimes kill wolverines over carcasses. To thrive, wolverines must be relatively fearless and willing to take more risks than the other carnivores.[17]

Food availability determines wolverine population density. North American wolverine density estimates vary from one per 25 square miles in Montana, in national parks where there's no trapping, to one per 77 square miles in northern British Columbia, Alaska, and the Northwest Territories, where trapping is allowed. In the far north, wolverines can be so dependent on carrion left by wolves that their density and conservation status are linked to wolf density. And where wolves have been removed by intensive management, wolverine numbers decline.[18]

Because of wolverines' generalist (i.e., broad) feeding habits, early researchers Maurice Hornocker and Howard Hash called them marginal carnivores on the fringe of the main links of the food web. In other words, because they primarily scavenge, wolverines don't drive trophic cascades the way wolves do. Yet despite the fact that they don't drive any trophic cascades that we know of, their predation and scavenging helps energy cycle through ecosystems.[19]

Wolverine Conservation History

During our Carnivore Way journey, we stopped to get gas at Haines Junction, Yukon (pop. 589), a village with a strong aboriginal culture. As I paid for the gas, I spotted a wolverine pelt hanging on the wall behind the cash register. The handsome, full-size pelt was mounted on a red felt backing. The species' characteristic diamond of tan-colored fur stood out against the rich mocha of the rest of the animal's body. A white price tag marked $500.00 dangled from one corner of the backing. Since reaching the Yukon two weeks earlier, I'd been observing wolverine pelts in at least half of the gas stations and convenience stores where we stopped. The pelts provided a revealing glimpse of the local culture and a context for carnivore conservation in this outback. They eloquently illustrated this species' importance as a trade item in the Yukon's subsistence culture. (However, humans don't typically eat wolverine meat, because of its rank taste.)

To understand wolverine conservation history, it helps to begin by learning about fur trapping—a human tradition that has and continues to shape wolverine conservation. In cold climates, humans have always used fur to stay warm in winter. Cave pictograms that date back to 3500 BCE provide evidence of humans trapping animals for their fur. And today in Alaska and the Yukon, the Inuit trim their parka hoods with wolverine fur because of its frost-resistance.[20]

Wildlife managers classify wolverines as *furbearers*—species traditionally trapped or hunted primarily for their pelts. A diverse group of animals, furbearers include both carnivores and rodents, such as beavers. A dense, soft, insulating underfur and an outer layer of long, glossy guard hairs characterize furbearer pelage.

Historically, the value of wolverine pelts has influenced this species' conservation. As European settlement progressed in North America, fur became the principal article of commerce. More recently, subsistence living continues to be an important part of the far north's culture. This includes trapping furbearers to bring in income—not just provide food or shelter. Today, many aboriginal communities in Alaska and northern Canada continue to rely economically on the sale of pelts. British Columbia has approximately 2,900 registered traplines and considers furbearer harvest a major source of income. Other animals often trapped include beaver, marten (*Martes americana*), bobcat (*Lynx rufus*), lynx, coyote (*Canis latrans*), and wolf.[21]

Scientists have taken a close look at the ecological sustainability of wolverine trapping. In areas with highly regulated trapping, such as Montana, Inman and others found that wolverine offtake had a compensatory effect on mortality. A very low trapping rate reduced wolverine numbers to the same level to which they would have reduced themselves via natural mortality (e.g., wolverines starving to death, or males killing other males in territorial disputes). Here, Inman and colleagues found a sustainable wolverine trapping quota of less than 6 percent of the population. Because smaller, isolated mountain ranges were more susceptible to wolverine over-harvest, they recommended smaller trapping-management units to reduce the probability of local declines. In British Columbia, which has a relatively high wolverine population, Eric Lofroth and Peter Ott found sustainable wolverine trapping in 56 out of 71 wolverine-management units. However, further research is needed to prove that trapping is anything but additive for a species that has such low reproduction and survival rates.[22]

One of the leading problems with trapping is that traps set for other species kill many unintended victims, including wolverines. Not always reported and difficult to manage, such losses can be devastating ecologically (e.g., when a breeding female is lost). For example, in northeastern Oregon, using camera traps in an area with no previous records of wolverines, Audrey Magoun documented three individuals of this species in 2011–2012. In December 2012, a fur trapper inadvertently caught one of them in a bobcat trap. He reported the incident to the Oregon Department of Fish and Wildlife and released the wolverine unharmed. However, this incident illustrates how trapping can affect non-target species.

Opposition to fur trapping arose in the early twentieth century, led by animal rights activists in response to the near extinction of furbearing species. As a result, since the 1950s, provincial, state, and federal laws have regulated trapping. However, regulation and enforcement vary widely. Today most people see trapping as an anachronistic, unacceptable activity. While trapping has died out in most urbanized parts of North America, it remains active in the far north and in rural places with high furbearer populations.

Some consider trapping a way to hold onto wilderness values and traditions, such as living off the land, that have long defined us as humans. The Wildlife Society (TWS), which supports science-based wildlife management, endorses trapping as "part of our cultural heritage that provides income, recreation, and an outdoor lifestyle for many citizens through use of a renewable natural resource." Montana Fish, Wildlife, and Parks calls trapping "a time honored heritage," one that is "biologically sustainable and is an important part of Montana's cultural history and outdoor lifestyle." Under the aegis of wildlife management, some see trapping as an effective way to manage problem animals, such as raccoons (*Procyon lotor*).[23]

Several problems exist with this line of thinking in support of trapping. First, as to its being a traditional subsistence endeavor, trapping as practiced today hardly resembles the trapping of the past. Modern trapping methods make furbearers more vulnerable than ever before. Rather than using snowshoes and dogsleds to check traplines, today's trappers use snowmobiles that can run at high speeds and for long distances. This enables them to cover more ground, expand trapline length and the number of traps, and go deeper and higher into wolverine habitat. Second, trapping often involves a slow death for animals, and as such it can hardly be regarded as ethical sport. Third, due to its somewhat indiscriminate nature, in areas with threatened or endangered populations of species, such as wolverines, trapping simply isn't biologically sustainable. Finally, that trapping is a traditional activity doesn't automatically make it ethically defensible today. For example, most cultures once practiced slavery, yet no longer condone it.[24]

Despite changing attitudes, some people continue to trap as their primary source of income. My northwest Montana writing cabin lies in the northern

Rocky Mountain foothills, in a place that gets six feet of winter snow and has one of the largest populations of furbearers south of the US-Canada border. A handful of people live in this remote, extreme area, accessible only via unpaved Forest Service roads that often become impassable in winter. In the 1970s, some came here to escape civilization and engage in a mountain-man subsistence culture. One of my neighbors was among those who came in the 1970s—and stayed. He's practiced subsistence living since then, supporting a household and raising a family with proceeds from his trapline. He lives simply and takes pride in his work. Nevertheless, electing to make one's living by trapping raises the ethical issues outlined above.

We've gleaned what we know about the wolverine's historical range from trapping records, expedition accounts, and museum specimens. In 1994 and 2000, the US Fish and Wildlife Service (USFWS) considered listing the wolverine under the Endangered Species Act (ESA). Each time, USFWS found listing unwarranted due to a lack of reliable information about this species' historic range. To address this lack, Keith Aubry and colleagues conducted a comprehensive review of wolverine trapping and observation records. Between 1801 and 1960, wolverines occurred in the northern and southern Rocky Mountains, the Washington and Oregon Cascade Mountains, the California Sierra Nevada, and the Upper Midwest. Between 1961 and 1994, people reported wolverines in Montana, Idaho, and the Cascades, with one verifiable wolverine report recorded for Wyoming and one for Nevada. Between 1995 and 2005, wolverines declined significantly in the Cascades and Rockies. However, in 2008, USFWS deemed putting the wolverine on the list of threatened and endangered plant and animal species unwarranted because none of the above regions contained a Distinct Population Segment (DPS). Earthjustice, an environmental law firm, sued, alleging violation of the ESA.[25]

Currently, 250–300 wolverines live in the contiguous United States. Figure 6.1 depicts wolverine distribution in 2013 within the Carnivore Way. The most stable population occurs in the Northern Rockies. Scientists have attributed this population's ongoing decline to climate change, which is ineluctably reducing wolverine range to isolated mountaintops—creating habitat islands that impair this species' ability to maintain its genetic diversity. Other threats

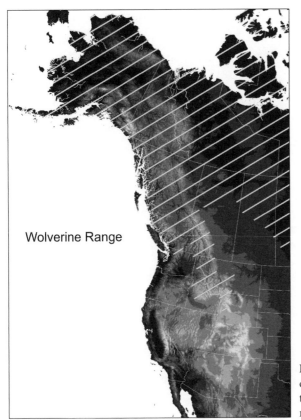

Wolverine Range

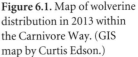

Figure 6.1. Map of wolverine distribution in 2013 within the Carnivore Way. (GIS map by Curtis Edson.)

include impacts from backcountry recreation, roads, and trapping.[26] By 2010, wolverine trapping was prohibited in all of the United States, except for Alaska and Montana. In October 2012, environmental groups litigated the ecological soundness of lethal wolverine trapping in Montana and prevailed. Meanwhile, wolverine numbers continued to drop.

In December 2010, USFWS designated the wolverine a candidate for ESA listing. However, a backlog of other candidate species effectively put the wolverine at the back of the line. Concern arose that given this species' low reproductive rate and rapidly shrinking habitat, it might go extinct even before USFWS was able to list it. When another lawsuit propelled the wolverine to the front of the line, USFWS agreed to submit a proposed listing rule by 2013.[27]

Even without all the litigation back and forth, listing a species under the ESA is a cumbersome bureaucratic process at best. It took nearly two years for USFWS to file a petition to give "threatened" status to the wolverine in the contiguous United States. This petition paved the way for a wolverine reintroduction first proposed by Colorado in 2010. The Service is proposing a 10(j) designation for the species in Colorado and portions of Wyoming and New Mexico.

USFWS maintains that climate change and trapping present the wolverine's primary threats. The agency doesn't consider the use of snowmobiles and timber harvest to be threats, because these activities occur on a small scale. This may not be a valid assumption, because as the climate continues to warm and snow cover shrinks, both wolverines and humans will be concentrating their presence and activity in areas with snow to smaller and smaller patches, where wolverines will feel the negative impacts of humans more strongly. The 10(j) rule allows for incidental killing of wolverines during otherwise legal activities, such as fur trapping. The rationale for this is that while in Colorado, state regulations prohibit trapping, beyond this state USFWS considers the portions of Wyoming and New Mexico where incidental trapping could occur to be such a small proportion of the nonessential experimental population area as to present a negligible threat.[28]

In the far north and Canada, the wolverine has never been extirpated. Western Canada has an estimated wolverine population of 15,000 to 19,000 animals. These numbers have been arrived at via extrapolation from small, known populations, so they represent an educated scientific guess. While this sounds like a lot, a closer look identifies weak links in this species' Canadian conservation status. Lofroth and Krebs projected a British Columbia wolverine population of 3,530. Considered vulnerable to habitat changes and threats, the wolverine has blue-list status there under provincial law. And on Vancouver Island, the wolverine has provincial red-list ("endangered") status, because humans extirpated it. In Alberta, wolverine trapping has declined steadily over the last two decades, which suggests that the wolverine population is declining as well. In recognition of threats to the wolverine and lack of information, Alberta has designated it a "data deficient" species. To fill this information gap, the province formed the Alberta Wolverine Working Group. In the meantime, wolverine trapping continues. On a federal level, the Canadian Committee on the Status of Endangered Wildlife (COSEWIC) categorized the wolverine as a "species of concern."[29]

The International Union for the Conservation of Nature (IUCN) has designated the wolverine globally as "of least concern," due to its wide distribution and large populations in places such as Alaska. IUCN scientists have identified a global decline of this species due to trapping and land use (e.g., roads, snowmobile recreation), but they don't consider this decline sufficient to warrant listing. Nevertheless, as climate change progresses, the IUCN may be re-classifying the wolverine as "vulnerable."[30]

Wolverine Habitat Needs and Climate Change

Snow availability is the single most important wolverine habitat need. In the far north, because of the deep cold, wolverines can use a variety of habitat. Farther south, wolverine habitat occurs only in the mountains. Because of their dependence on snow, in the contiguous United States wolverines exist as small, demographically vulnerable populations.[31]

The Intergovernmental Panel on Climate Change (IPCC) defines *climate* as weather conditions over time, with 30 years being a typical period for such measurements. *Climate change* refers to a change in the mean or variability of one or more measures of climate (e.g., precipitation, temperature) over an extended period of time, typically decades. Since the 1980s, scientists have recognized that our world is warming. A warming world will have less snow.[32]

Several recent studies have modeled the expected direct effects of climate change on wolverines. All found wolverine presence linked closely to late-spring snow abundance and that climate change would reduce wolverine habitat and corridors. For example, Copeland and colleagues used models of snowpack to evaluate the locations of known wolverine den sites globally. They found a 95-percent association between snow cover and wolverine summer presence. Jedediah Brodie and Eric Post used trapping success as an indicator of wolverine population. They found a correlation between the decline of trapping success and the decline of snow cover.[33]

Looking toward the future, McKelvey and his colleagues examined the effects of climate change on wolverines over the next century. Their models predicted declines in snow cover and wolverine habitat from 2030 to 2099. Areas of extensive contiguous spring snow cover will continue in British Columbia, north-central Washington, northwest Montana, the GYA, and Colorado. However, with 21 peaks with summits above 14,000 feet, Colorado will have the most enduring wolverine habitat. By the late twenty-first century, increasingly isolated wolverine habitat will lead to genetic isolation as it becomes more difficult for wolverines to travel from one habitat patch to another. By the twenty-second century, these smaller, more isolated populations may face extinction.

Gene flow among wolverine populations in Canada, Glacier National Park, the Idaho Panhandle, and the GYA is essential for survival of this species in the contiguous United States beyond this century. Random changes in the gene combinations of a species, called *genetic drift*, can result in loss of favorable genetic traits and endurance of traits that randomly show up and are biologically unfavorable. Genetic drift can occur only in isolated populations with such a small gene pool that random events change their makeup substantially. In a genetic study of US wolverines, Chris Cegleski and colleagues found that this species' persistence will depend on female movement into new territories. To ensure genetic diversity, they recommended maintaining at least 400 breeding pairs in the United States in the wild, or effective migration of one to two female wolverines per generation.[34]

Those of us who live in the Northern Rockies and the far north have seen climate change develop far more rapidly than any of the models that existed in 2007 predicted. The speed with which such change is progressing warrants a precautionary approach. Indeed, planning for a worst-case climate scenario may be the wisest wolverine-conservation strategy.

Wolverines and Corridor Ecology

Due to climate change, perhaps no other species is more emblematic of corridor ecology than the wolverine. Given this species' elusiveness, we were unable to learn about its movements and full habitat needs (e.g., corridors) until radio collars arrived on the scene.

Early wolverine research began in the late 1980s with work by Hornocker and Hash in Montana's Swan Range, by Audrey Magoun in Alaska, and by Vivian Banci in the Yukon. In the 1990s, Krebs and Lofroth had just begun research in British Columbia. Back then, objectives were simple: scientists wanted to learn more about this mysterious animal's ecology. However, due to the challenging subalpine and alpine terrain in which the wolverine lives, learning the bare bones of its ecology required an epic trial-and-error effort.

Wildlife biologist Jeffrey Copeland began to study wolverines in 1992, in Idaho's Sawtooth Mountains, as a game warden turned graduate student. He and other early researchers began with wolverine basics—pioneering the best way to trap and radio-tag these collar Houdinis and keep up with them once they've been successfully tagged. In the process, researchers learned important things about this species' ecology, such as their preferred habitat, home range size, and reproductive behavior.

Using radio-collar data, Copeland built on earlier wolverine studies. He fearlessly followed wolverines through territories as large as 200 square miles in the Frank Church River of the No Return Wilderness. When funds allowed, he monitored collared wolverines by aircraft. But typically his work involved radio-tracking them on foot, using a receiver and antenna through terrain nearly impassable at times. In his study, Copeland learned that young male wolverines make very big movements and use corridors. He began to learn about surprisingly complex wolverine social interactions, as described above in the natural history section of this chapter. For example, he documented a father teaching a daughter to hunt. He also found evidence of a mother burying one of her kits, who had died after being implanted with an abdominal transmitter. Finding that kit's grave had a profound effect on Copeland as a scientist and made him consider more fully the impacts of wolverine research on the individuals that he was handling. He kept this Idaho wolverine study going until 1996 and became the director of the Wolverine Foundation.[35]

Since 2001, Inman's WCS Greater Yellowstone Wolverine Program has taken a landscape-scale approach to studying this species' ecology. He and his colleagues have been using wolverine GPS-collar data to map this population's habitat needs and corridors and to find potential climate-change refuges.

They're further identifying essential wolverine habitat linkages in order to develop a connectivity conservation strategy.

According to Inman, the GYA wolverine population represents a metapopulation isolated from others. Thus, its persistence depends on successful dispersal. Additionally, the patchy nature of wolverine habitat in the GYA
creates isolated demes (defined as a local population of a species) within this
region's metapopulation (defined as a population of populations). In 2008, Inman found that the southwest Montana mountains contain three adult male
and six adult female territories. These demes make up a metapopulation whose
viability (i.e., genetic diversity and persistence over time) depends on successful
wolverine dispersal from Idaho and Montana.[36]

Because of wolverines' crucial need for successful dispersal, Inman suggests thinking beyond the traditionally observed boundaries of the Crown of
the Continent Ecosystem and GYA. Potential dispersal corridors lie where these
areas overlap. This holistic view of the US Northern Rockies yields fourteen
wolverine demes. Only four have the potential to sustain substantial wolverine
populations (i.e., more than 50 animals), which means that they function as
population cores (defined as a thriving deme of a species living in high quality
habitat). Most lie on public lands. However, for wolverines to disperse between
demes successfully, the in-between lands—the matrix—must be managed in a
dispersal-friendly manner. Securing connectivity for these demes will require
collaboration with local communities, government agencies, and scientists.[37]

As of 2009, the 40 adult female wolverines Inman had monitored had
taught him a great deal. He learned that they had an average reproductive rate
of 0.24 kits per female per year, or approximately one kit every four years. Adult
mortality was low and attributable to natural causes. They primarily ate ungulates in winter and marmots in summer. And dispersing wolverines sometimes
crossed major highways. Consequently, Inman recommended highway mitigation to maintain dispersal linkages.[38]

For his doctoral studies, Inman defined the wolverine's ecological niche by
analyzing its movements and ecology in the GYA. He found that wolverines
were limited to high elevations with low temperatures, abundant winter snow,

and complex landscape structures, such as downed trees and boulder talus slopes. Survival required large home ranges and caching food during all times of year. He found caching may be crucial for the reproductive success of this species. He estimated the carrying capacity of wolverine habitat south of the US-Canada border at 580 individuals and the current population at approximately half of this. To address the effects of climate change and increasing human development, Inman recommended: 1) maintaining connectivity, particularly in the US Northern Rockies; 2) restoring wolverines to areas of historical distribution more resistant to climate change, such as Colorado peaks; and 3) developing a collaborative multi-state/province monitoring program for this species.[39]

From 2002 through 2008, Copeland and Yates led the Glacier National Park Wolverine Research Project. Glacier is a key study area, because linkages still exist between it and Canadian wolverine populations. The primary research objectives were to radio-tag and monitor wolverines in order to describe their distribution and reproductive status, about which little was known in this park. Yates directed the field crew, which included Marci Johnson, veterinarian Dan Savage, Doug Chadwick, and others. Over five years, the research team captured 28 wolverines and fitted them with GPS collars or VHF abdominal implants.

Yates and his crew's field effort focused on the eastern portion of Glacier National Park, which contains the majority of wolverine habitat in the park. Across most of the year they radio-tracked wolverines, sometimes going into virtually inaccessible places where they endured subzero temperatures, blizzards, and avalanche hazards. The radio-tags that wolverines had (GPS collars and abdominal implants) produced 30,000 fixes. The research team mapped these fixes, which enabled them to locate 23 natal dens and 30 rendezvous sites. As in other studies, they found dens associated with deep snow. Many dens lay in areas with downed whitebark pine (*Pinus albicaulis*) or talus slopes, which provided structure for tunnels.

Some of Copeland and Yates's most pertinent findings had to do with wolverine mortality. They found low kit survival to adulthood. For example, five out of seven kits identified during the first year of research died. Causes included predation, falling off cliffs, and shooting. Adult and subadult mortality during the project reduced this population's viability. Two adults died of natural causes (avalanche, predation) and three subadults died due to legal trapping outside the park, road construction inside the park, and being gored by a goat.

After five years of study, Copeland and Yates recommended moderating human use of the park and surrounding areas in winter and especially in late spring, in order to protect denning habitat. The park allows most of its roads to become snowed-in during winter. However, spring plowing can disturb wolverines. Because the main road through the park lies within the home range of at least three wolverines tracked in this study, Copeland and Yates suggested adjusting plowing activity.[40]

Park administrators ended the study in 2008. In 2009, park ecologist John Waller initiated a wolverine DNA project. This new project involved collecting wolverine genetic data via hair samples. Waller and his technicians placed hair traps along wolverine travel corridors. Set into ice, often at frozen lakes, the traps consisted of posts baited with deer legs, with wire positioned to collect hair. Biologists and volunteers skied into some of the park's most remote recesses to retrieve the resulting hair samples. The study continued for three years, through winter 2011. Data from this project are still being analyzed.

In Canada, Tony Clevenger has combined citizen science with corridor ecology and population modeling to help conserve wolverines. He heads the Wolverine Watch Project, part of the larger Banff Wildlife Crossings Project. Clevenger's 2,320-square-mile study area lies in the Canadian Rocky Mountain national parks (which include the contiguous Banff, Yoho, and Kootenay National Parks), along the Continental Divide in British Columbia and Alberta. Functioning like a genetic funnel, this corridor connects northern wolverine populations to southern ones. Or at least it did until human development brought the four-lane Trans-Canada Highway (Highway 1) to Banff, along with millions of human visitors per year. Clevenger is focusing on identifying key wolverine travel corridors. This information will be used to design highway mitigation (i.e., crossing structures), as Highway 1 continues to be widened through prime wolverine habitat in the Continental Divide. His wolverine study will also be used to develop occupancy models in order to estimate presence and population size.

For this landscape-scale project, Clevenger gridded his study area into roughly 50 cells. Within each cell he created a hair trap made up of a whole, skinned beaver carcass nailed to a tree and secured with wire. Wolverines climb the tree to get the beaver meat, in the process leaving a hair sample on the wire. Remote thermal infrared cameras (for night detection) at each hair-trap site record activity. Figure 6.2 shows a wolverine caught in the act of raiding one of these traps. Teams of two—a professional biologist paired with a volunteer— check the traps regularly. To do this work, Clevenger has recruited volunteers from local outfitters, ski guides, climbers, and conservationists with backcountry ski experience.

Two years into this study, Clevenger has identified 22 individual wolverines via DNA analysis: fifteen males and seven females. So far only three individuals have been detected on both sides of Highway 1, all males and all crossing right at the Continental Divide. He's documented ten wolverine crossings of

Figure 6.2. Wolverine at one of the Banff Wildlife Crossings Project hair-snare traps in Banff National Park, Alberta. (Remote Camera Image, Anthony Clevenger.)

Highway 1 at wildlife crossing structures. But his most surprising finding has to do with wolverine abundance and distribution in national parks compared to neighboring Kananaskis Country (where hair traps have also been established). He found that wolverines visited nearly 90 percent of all hair traps in the national parks, but only 25 percent of the Kananaskis hair traps. This shows that the contiguous Canadian national parks (Banff, Yoho, and Kootenay) are clearly an important core area for wolverines in the Rocky Mountains, and the area closest to the critical core in Glacier National Park, Montana. Clevenger has now turned to looking at wolverine corridor ecology in the portion of the Crown of the Continent Ecosystem that links Glacier to Banff.[41]

Coexisting with Wolverines

Coexisting with wolverines requires respect and awareness of their needs. While wolverines don't threaten humans and don't eat livestock, coexistence has to do with finding ways to moderate our use of resources (e.g., trapping, backcountry recreation, timber harvest). With the added pressure of climate change, enabling wolverines to thrive in a human-dominated world means finding creative solutions. A wolverine reintroduction in Colorado (with its high peaks) has emerged as one of the more promising ways to conserve this species. Inman is among those leading the reintroduction, supported in this by the body of scientific literature on wolverine conservation needs and the recent USFWS proposal to put this species on the list of threatened and endangered plants and animals.

All the science to date clearly shows that to live long and prosper as a species, wolverines need a network of secure habitat and connected corridors that link populations and the precious genes they carry. In this chapter we saw how scientists have been identifying these travel routes more clearly in order to enable wolverines to move securely from one chain of peaks to another, and also filling gaps in our knowledge of this species' biological needs. In our rapidly warming world, it will ultimately take a combination of government protection, reintroduction efforts in places like Colorado, protected corridors, and changes in natural-resources management if we are to maintain this species. National parks provide wolverine refuges. But parks alone are little more than postage stamps of security for a species that needs so much room to roam. As Chadwick puts it, "If the living systems we choose to protect aren't large and strong and interconnected, then we aren't really conserving them [wolverines]. Not for the long term. Not with some real teeth in the scenery. We're just talking about saving nature while we settle for something less wild."[42]

Lynx (*Lynx Canadensis*). (Drawing by Lael Gray.)

CHAPTER SEVEN

Lynx (*Lynx canadensis*)

In the late 1990s, my husband Steve and I spent a year planning the construction of a writing cabin for me in a remote corner of our Montana land, where we live. While all I wanted was a modest cabin, the project quickly became complicated and expensive. This prompted us to consider other options. We checked out real estate listings and found decaying cabins jostled cheek by jowl next to similar cabins on third-of-an-acre scraps of land, priced well into the six figures. That wasn't what we wanted or could afford.

We'd almost given up on making my writing cabin a reality when, one morning over coffee, Steve found a promising ad in the Mountain Trader—a weekly newspaper that advertises horses for sale next to ads for shotguns, firewood, and used baby furniture. The ad read, "TEN ACRES: Log home & guest cabin, creek, mountain views, borders Forest Service. Cash!"

Steve called right away, and it sounded too good to be true. Not one cabin, but two, and one of them the original homestead cabin to boot; not one stream, but three—two of them teeming with native cutthroat trout. The place had abundant wildlife and end-of-the-road privacy—all for a very low price. Drawbacks included no electricity or plumbing (though there was a deluxe two-seater outhouse), and the fact that the cabins were still "under construction." Even so, it sounded like a great deal, so the following morning we went to look at the property.

We drove north toward the Canadian border and then west into the mountains, to an area where only 1 percent of the land is privately owned and the rest is national forest. The minute we entered the original plain-built homestead cabin, it felt like home. It needed work, but there was nothing that couldn't be fixed with elbow grease.

Our deciding moment came as we stood outside, deliberating the pros and cons of this purchase. All at once, three animals bounded across the sparsely forested slope between the cabin and the first stream: a mother lynx with two half-grown kittens. When they stopped to wrestle with each other, their mother cuffed them gently and moved them toward a willow thicket. I'd never seen lynx in the wild before and knew the US Fish and Wildlife Service (USFWS) had just put them on the list of threatened and endangered species of plants and animals. Moved by this intimate look at wildness, we struck a deal with the owner on the spot.

We took possession two months later. Among the closing papers, we found a deed stating that on May 13, 1968, a Washington logging company sold the property to "Printer L. Bowler (a single man)" for the astonishing sum of "$10.00 and other good and valuable consideration." Two years later, Printer sold the property to a hermit, a single man, who built a cabin from lodgepole pines logged from this land. Thirty years later, a married couple bought the property from the by-then-elderly hermit. She was a forty-something, fine-boned beauty named Montana. He was a darkly handsome drifter from another state, a mountain man wannabe two decades her junior, whom she met in a bar. Four years later, when their marriage fell apart, Montana put the property on the market.

While we waited for escrow to close, Montana walked the land with us and showed us the three creeks on it, one without a name. She showed us secret places, too: where to find huckleberries in August and the marsh where moose calved in late spring. We signed the final papers in a lawyer's office amid hugs and bittersweet tears.

We got to work right away. We plugged gaps between logs that had been stuffed with rags to keep out the cold, replaced the chinking, windows, and doors, and installed plumbing and electricity. Once we leveled the subfloor, with a carpenter's help we put in a beautiful wormwood pine floor, using ecologically sustainable salvage wood from a friend's small mill. After the cabin was livable, we filled it with comfortable furniture, books, dogs, children, friends, and good love. We named it Moosewillow for the moose and willows that abound here.

Moosewillow has no street address or mail service. The dirt roads that lead here are mostly unmarked. Nobody can find us unless we wish to be found, which is fine with us. Quite simply, it's paradise. It lies in a drainage that contains one of the richest wetlands in northwest Montana. In summer, cutthroat trout leap in the creeks. In autumn, an elk herd passes through, the bulls chasing the cows, bugling and pawing the ground, tearing up the trees with their antlers. Cougars and wolves follow the elk. Three wolf packs range in the area surrounding our property. In winter, curious moose fog our cabin windows with their breath, and friends come up for cross-country ski weekends. In fact, we love spending time there so much that we often spend winters there, and the rest of the year at our regular home.

We discovered an amazing winter world at Moosewillow. Our land lies in a snowbelt where winter snow depth exceeds six feet and the mercury plummets to minus-50 degrees Fahrenheit. From our land you can ski for hundreds of miles on old Forest Service roads and game trails. We've learned how to break our own ski trails using short, wide backcountry skis. As we ski, sometimes we find live bear dens, given away by the occupant's warm ursine breath emerging from an opening in a mound of snow, condensing into great clouds that hang heavily in the frigid air before dissipating. And on these winter forays, we used to find lynx stories in the snow.

One of the more common and compelling winter stories would begin with the elliptical tracks of a snowshoe hare as it hopped from shrub to shrub, foraging. On each outing, we'd find hundreds of such tracks in the snow. Inevitably we'd also find a few smooth, soft-edged depressions under the fir boughs that overhung the trail's edges. These depressions marked the spots where a lynx had sat on its haunches, melting the snow slightly as it waited for prey. A few feet from these ambush beds we'd find the place where a snowshoe hare had come by, and the lynx had made its move. In one clean leap, the lynx had taken the hare by surprise, and in another move killed it by breaking its neck, leaving a crimson blossom of blood on the snow. We'd come upon these snow stories nearly every time we went skiing. However, while we can still ski for miles and miles around Moosewillow, now we rarely find such signs, because the hares and lynx are gone.[1]

About five years after we bought Moosewillow, the forest that surrounded our cabin was logged. Before the cut, this forest consisted of a dense doghair thicket of lodgepole pines intermingled with old cedars and larches. This unusual tree combination had resulted from a fire 80 years earlier that had left a massive amount of lodgepole pine seeds. Lodgepole cones are partially *serotinous*, which means that they can open and release some of their seeds only during a fire. This fire had created a seed-rain event, in which the wind had picked up seeds from the open serotinous cones and deposited them in the area around Moosewillow when the wind died. Due to subsequent fire suppression, 80 years later the lodgepole pine density had reached nearly 1,000 trees per acre. This overgrown forest needed to be thinned in order to mimic the ecological effects of wildfire. We knew about the planned thinning when we bought Moosewillow.

The forest plan called for thinning using modest *aggregated retention*, which would protect and improve lynx habitat. This optimal thinning would mean retaining clumps of old forest and creating small openings. In the 1990s, forest ecologists like Doug Maguire developed this method of timber harvest to maintain wildlife habitat and biodiversity on managed forest lands. Twenty years later, forest ecologist Jerry Franklin and policy expert K. Norman Johnson recommend aggregated retention as one of the best ways to create healthy working forests.[2] Science furnished ample evidence that lynx avoid clearcuts and sites that have been heavily thinned. They prefer habitat "mosaics" that provide abundant cover and diverse structure, such as deadfall and shrub thickets.[3] When I saw the plan for the forest around Moosewillow, I'd just begun my master's studies in conservation biology and knew little about forestry. To my inexperienced eyes, the plan looked good. In a system lacking fire, the thinning could create the sort of habitat mosaic preferred by lynx. Now that I have a PhD in Forestry and Wildlife, the original plan still looks pretty good. But sometimes what's implemented on the ground out in the back of nowhere isn't exactly what's planned officially.

Late one winter night I came home to Moosewillow (where we had been living that winter) from a research trip to find a moose standing in the middle of the road. The moose began to trot in front of my car as I drove through the logging debris. Even in the dark it looked bad—broken trees everywhere, the air redolent of bruised pine. The next morning I awoke to witness a pair of robotic logging machines cut down the last trees in front of our cabin, pincher arms snapping tree trunks in two like twigs, giant tractor-tread tires crushing everything in their path. The earth shook as the machines lumbered through what was left of the forest, reminding me of the T. rex scenes in the *Jurassic Park* films. When all was said and done, this commercial thinning resulted in a forest canopy of less than 5 percent. To enable the remaining trees (all larches, which have high economic value) to grow without competition for future harvest, the logging company had sprayed herbicide, thereby preventing shrubs from sprouting for many years to come.

After the cut, Forest Service personnel hosted a community meeting in which they shared their desire to maintain forest health and save cabins such as ours from the risk of catastrophic fire. Never mind that many of us who owned places like Moosewillow accepted the fire risk that came with such choices. Forest Service ecologists explained the various degrees of logging operations, or "treatments." I learned that what had happened at Moosewillow was a heavy treatment. The logger sitting beside me grinned and whispered, "The next best thing to a clearcut." A clearcut was proposed for an adjacent forest patch. Later, when I reviewed the timber sale documents, I learned that the cut around Moosewillow and the next planned cut both failed to meet the federal guidelines established in the Canada Lynx Conservation Assessment published in the year 2000, and in the Forest Service's own 2004 Northern Rockies Lynx Environmental Impact Statement. A total of 23 miles of road would be built near our land in order to access additional harvest units. The effect of these new roads would be mitigated by the closure of other, pre-existing roads, and in some cases by limiting seasonal access. The new roads would nevertheless negatively impact lynx, which, like most carnivores, avoid roads because they bring humans.[4]

I was so troubled by the logging around Moosewillow that I finished my master's degree in conservation biology and immediately began a PhD in Forestry and Wildlife. Since then, I've learned enough to realize that what happened at Moosewillow was an exception rather than the norm in this era of ecosystem management, which calls for managing forests for the benefit of whole ecological communities.[5]

Now it's been more than a decade, long enough to assess the longer-term effects of this cut. The lynx that abounded at Moosewillow left after the timber harvest. They haven't returned, as their habitat is gone. It'll take a couple more decades for the forest to grow back to the point that lynx will have the sort of complex, multistoried forest with the dense understory that they require for hunting and denning. Snowshoe hares are uncommon in the area, as they lack the shrub cover that was removed as a result of the harvest. Other species have been affected as well. Like lynx and snowshoe hares, moose prefer multistoried forests. And like lynx, they left after the timber harvest and still haven't returned. However, wolves are doing fine, as are bears, and of course deer and elk, which prefer open forests.

Since the logging at Moosewillow, there's been additional timber harvest in the area. Most of these timber sales have been conservative in terms of the number of trees removed, and they have adhered to lynx-management guidelines. However, in revisiting lynx-conservation issues today, the subject of logging keeps coming up and remains unresolved. To understand lynx habitat needs, the best place to begin is this species' taxonomy and fascinating natural history.

What's in a Name: The Seer from Canada

The lynx (*Lynx canadensis*) is a midsized New World member of Felidae, the cat family. Taxonomists recognize two North American subspecies: *L. canadensis canadensis* in the northern United States and Canada, and *L. c. subsolanus* in Newfoundland. *L. c. canadensis* is the only subspecies within the Carnivore Way. Its genus name, *Lynx*, which comes from the Greek word for "lamp"—a reference to its lucent eyes and keen vision. Its species and subspecies names mean "of Canada" and have to do with the fact that it was first documented there.[6]

Lynx Natural History

While I was visiting carnivore ecologist John Weaver at his Montana home, he showed me the complete skeleton of an animal mounted inside a Plexiglass case and asked me to identify the species. Smallish and fine-boned, the animal's hindquarters and hind legs were far longer and more robust than its shoulders and front legs. It had impossibly large feet.

"Snowshoe hare?" I guessed, before looking at its head.

"Look again," said Weaver.

Its skull had the unmistakable sharp teeth of a carnivore. Its blunt snout let me know it wasn't a member of the dog family.

"Lynx," John said. He went on to explain how lynx body shape (e.g., big feet and long, powerful hind limbs) resembles that of its primary prey, the snowshoe hare (*Lepus americanus*). In this remarkable example of evolutionary relationships, these two tightly coupled species had developed the same physical traits allowing them to move efficiently through the northern deep-forest habitat that they share.

Lynx are built for snow. Their broad, furry paws function like snowshoes, and their well-developed hindquarters enable them to move easily through deep snow. In addition to long legs and big feet, lynx have lean bodies and stubby tails. They have large, yellow eyes set into a square face, ears tipped with black hair tufts, and a prominent throat ruff. In winter, they have dense, grayish-brown to pale tan fur on their back, with silvery fur on their belly, legs, and feet. In summer, their fur turns tawny brown. People often mistake bobcats (*Lynx rufus*) for lynx. While the two species are similar in size, lynx have longer legs, bigger feet, more prominent ear tufts, and they lack the black markings that bobcats have.[7]

Lynx are considered *mesopredators*—midsized carnivores. They average 20 to 25 pounds in weight, roughly twice that of a domestic cat. Like other large carnivores, males are slightly larger than females. Male and female lynx have relatively large home ranges for their size, from 12 to 83 square miles. Home range size varies based on prey abundance, the density of the lynx population, and gender. Males have larger home ranges than females. Both sexes make long-distance dispersals to find mates and food, traveling as far as 2,000 miles, as we saw in chapter 1.

Lynx live in moist boreal forests. Snowy, wet winters and abundant snowshoe hares characterize these forests. Lynx habitat usually lies above 5,000 feet in lodgepole pine (*Pinus contorta*) or Engelmann spruce–subalpine fir (*Picea engelmannii–Abies lasiocarpa*) forests sufficiently mature to provide adequate cover from other predators, such as cougars (*Puma concolor*) and wolves (*Canis lupus*).[8]

Lynx reproduction resembles that of other North American wild cats. These mostly solitary creatures come together primarily to breed. They reach sexual maturity at the age of two and mate between February and March. They den from early May to mid-June in mature forests. Lynx select den sites based on structures that provide both cover and shelter. These structures include wind-felled trees, root-wads, or dense live vegetation. In the Yukon, lynx sometimes den in riparian willows, which offer thick, horizontal cover, and in old burned forests that have lots of deadfall and sprouted trees. They place their dens in forest swales (bowl-shaped depressions in the landscape) and avoid areas with a thin forest canopy (i.e., less than 50 percent cover) and within a quarter mile of roads.[9] After a two-month gestation, females give birth to two to four spotted kittens. However, females in poor physical condition have lower breeding success. Sometimes newborn kittens die soon, due to maternal malnutrition, which affects kitten body condition at birth as well as the ability of mothers to nurse their young. Females wean kittens by the age of three months. Kittens stay with their mother for a full year, and then establish their own territories.

Lynx primarily eat snowshoe hares. In fact, snowshoe hares make up to 97 percent of lynx diet. This inseparably binds lynx conservation to snowshoe hare conservation. While lynx also eat other small mammals such as red squirrels

(*Tamiascurus hudsonicus*), these alternative prey don't fully meet their nutritional needs. This is because squirrels are smaller than hares, and lynx must therefore expend more effort to catch enough squirrels (compared to hares) to nourish themselves.

Because lynx and snowshoe hares are so closely coupled via food web relationships, in order to understand lynx ecology we need to understand snowshoe hare natural history and ecology.[10]

Snowshoe hares live throughout northern North America and as far south as New Mexico. Their name comes from their large hind feet. These forest herbivores' primary habitat requirements are thick forest with a dense shrub understory and snow cover. Forest density, rather than age or specific understory-species composition, matters most. Conservation biologist Scott Mills and his colleagues found that open forests function as snowshoe hare population *sinks*—places with high snowshoe hare mortality due to predation. Conversely, thick, closed forests function as population *sources*—places where snowshoe hare populations can reach high densities. This means that young forests with thick understories that provide good cover for hiding from predators meet snowshoe hare habitat needs better than any specific forest-understory-species composition. Accordingly, snowshoe hares occupy diverse forest types, from Douglas fir (*Pseudotsuga menziesii*) to cedar (*Thuja* spp.), limber pine (*P. flexilis*), and juniper (*Juniperus communis*). They utilize a wide range of shrubs for food, eating mostly twigs and bark in winter, and leaves and berries in summer.[11]

For camouflage, snowshoe hares have two color phases, or *morphs*. In winter, they turn white to blend in with the snow. In summer, they turn medium brown, but their flanks remain white year-round. Their cryptic coloring makes it more difficult for predators to detect them. While lynx are snowshoe hares' main predator, other animals that prey on them include coyote (*Canis latrans*), bobcat, wolf, and cougar, as well as birds such as golden eagles (*Aquila chrysaetos*) and owls.

Because of the natural-history traits of the snowshow hare, lynx habitat is tied to snowshoe hare preferred habitat. Now let's take a broader look at lynx denning and hunting needs, which we discussed earlier, and how they correspond to snowshoe hare habitat. Lynx prefer mature forest with abundant deadfall and canopy cover for denning. They locate their den sites adjacent to prime snowshoe hare habitat: mature or young forest with a thick shrub understory. In order for lynx to thrive, they need to find sites where these habitats are paired.[12] Until it was logged, the forest that surrounded Moosewillow provided exactly that combination.

The lynx–snowshoe hare population cycle has been one of the most-studied ecological phenomena. In chapter 2 we looked at the food web relevance of this cycle. Since the 1920s, ecologists such as Charles Elton, Adolph Murie, and more recently, Stan Boutin and Charles Krebs, have observed that in Canadian and Alaskan boreal forests, lynx and snowshoe hares have nine- to eleven-year population cycles. As snowshoe hare numbers go up, lynx numbers go up, and as snowshoe hare numbers decline, lynx numbers decline, with a one- to two-year lag between cycles.[13] So how high are snowshoe hare population peaks? Their numbers can peak at densities as high as 3,900 hares per square mile. At high densities, this species exceeds the forest's carrying capacity.

In nature's economy, all that goes up must come down. From its peak, the snowshoe hare population goes into a downturn. As snowshoe hares start to deplete their food resources (e.g., forest shrubs), and as predation on them by lynx and other carnivores continues to increase, snowshoe hare numbers begin to decline. When snowshoe hare numbers are low, shrubs begin to recover. After a year or two at a low density, snowshoe hare numbers start to increase again, and the cycle repeats.

Lynx adjust their diet and ranging activity to correspond to snowshoe hare abundance. During snowshoe hare lows, lynx rely on alternative prey and carrion, which are insufficient to keep them nourished. Feeding on alternative prey causes lynx body fat reserves to drop, and they become more susceptible to starvation or predation by larger carnivores. Lynx numbers usually stay low for three to five years, beginning their recovery after hare numbers begin to recover. Scientists still don't fully understand these protracted lynx lows.[14]

In the United States, snowshoe hare populations don't exhibit such pronounced population swings as are seen in Canada and Alaska. Furthermore, in

Colorado, southern Idaho, and Washington, neither snowshoe hare nor lynx populations reach the high densities found in Canadian and Alaskan boreal forests. Consequently, many ecologists believe that the southern lynx population's ecology may differ significantly from that of the boreal lynx. For example, southern lynx rely more heavily on red squirrels and other prey at lower elevations. Additionally, lynx abundance may be more variable in the south. In southern forests where snowshoe hares are relatively scarce and patchily distributed, lynx have low reproduction. Because of all of these factors, John Weaver gives lynx a low resiliency rating, with southern populations less resilient than those in the north.[15]

Lynx Conservation History

From the 1800s to the early 1900s, lynx occurred widely in the far north and in the northern contiguous United States. They lived in areas characterized by cool, moist coniferous forests and at higher elevations in the West. But by the 1960s, this fur-bearing species' population had severely declined south of the US-Canada border due to unregulated trapping for pelts and loss of habitat via timber harvest and human development (e.g., road construction). This decline became so severe that lynx became extinct in large areas of their southern range. Meanwhile, lynx continued to thrive in the far north.[16]

In 2000, due to the ongoing threat of extinction, the federal government listed the lynx as "threatened" in portions of the United States, under the Endangered Species Act (ESA). Within the Carnivore Way, lynx have federal protection in Colorado, Idaho, Montana, Oregon, Utah, Washington, and Wyoming. In New Mexico, they're a candidate species for ESA listing. Lynx don't receive federal protection in Alaska, the Yukon, or the western Canadian provinces, due to their abundance there. Figure 7.1 depicts lynx distribution in 2013 within the Carnivore Way.

The federal government protected lynx due to concerns about lynx and snowshoe hare habitat conservation on federally managed lands. In the United States, timber harvest and human recreation are the principal land uses that

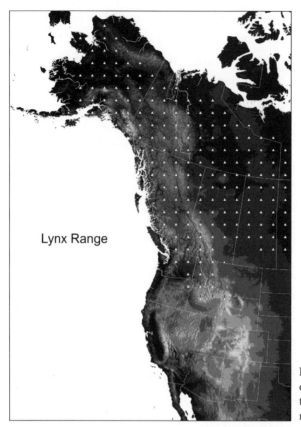

Lynx Range

Figure 7.1. Map of lynx distribution in 2013 within the Carnivore Way. (GIS map by Curtis Edson.)

negatively impact lynx habitat. Experts have expressed concerns about habitat fragmentation between lynx populations. For example, interstate highways bisect lynx habitat, creating the threat that lynx will die as roadkill. Wildlife crossing structures, though, have proven effective at maintaining lynx corridors. Figure 7.2 shows a lynx using a wildlife crossing structure in Banff.[17]

Fur trapping creates an additional threat. While, since 2000, lynx have been protected and can't be trapped below the US-Canada border, some states, such as Montana, allow trapping of other species. Between 1998 and 2006, nine lynx were reported caught inadvertently in Montana traplines. Although trappers attempted to release these animals, some died. Since no reliable population estimate exists for Montana lynx for this period, the effect of trapping those lynx on the overall population remains unknown. In several states, including Montana,

Figure 7.2. Lynx on an overpass of the Banff Wildlife Crossings Project Bow Valley in Banff National Park along Highway 1. (Remote Camera Image, Anthony Clevenger.)

environmental groups have filed lawsuits to halt trapping due to its perceived indiscriminate nature and the consequent threat it presents to lynx.[18]

Since giving the lynx ESA protection, the US federal government has been taking steps to recover this species. In 2009, USFWS defined and designated lynx critical habitat. However, thirteen years after listing the lynx, USFWS has yet to create a recovery plan. This delay has to do with lawsuits over lynx habitat in Washington, Idaho, Montana, and Wyoming (and also in the northeastern United States). For example, in 2009 snowmobile groups in Washington and Wyoming sued the federal government to nullify critical habitat designation in order to allow more snowmobile access. Earthjustice intervened, but some of these lawsuits have yet to reach closure. The delay in designating critical habitat also has to do with the bottleneck of other species in need of federal protection. Tired of waiting while the lynx's legal status remained unresolved, in 2013 environmental groups sued the federal government, demanding a recovery plan. In response to the lawsuit, USFWS issued a new definition of lynx critical habitat and a draft recovery plan.[19] South of the US-Canada border in the West, lynx numbers could be as low as several hundred. This gives urgency to finalizing and implementing the above federal documents.

To help increase the US lynx population, Colorado Parks and Wildlife (CPW) reintroduced lynx from 1999 to 2006. The core lynx reintroduction area lay in southwest Colorado, in Engelmann spruce and subalpine fir forests above 8,500 feet. Wildlife biologist Tanya Shenk led the reintroduction effort. Within the designated recovery area, over the years the CPW released a total of 218 lynx. The lynx were radio-collared, which enabled biologists to track their movements.

The reintroduction/augmentation phase of the project ended in 2006. At that time, most of the reintroduced lynx had remained near their release site. Eighty had died, with 31 percent of these mortalities caused by humans (e.g., through illegal shooting and collisions with vehicles). Starvation accounted for another 21 percent of deaths, and the rest were from unknown causes. By 2006, researchers had documented four cases of successful lynx reproduction.[20]

After 2006, the project focus shifted to monitoring the developing lynx population. Given this species' high starvation rate, the agency conducted a study of snowshoe hare densities, demographics, and seasonal movement patterns associated with different densities of forest cover from 2005 to 2009. CPW found adequate snowshoe hare densities to support lynx, and based on radio-collar data, 3.5 million acres of suitable lynx habitat in Colorado. This habitat occurred primarily in the southwestern part of the state, but also farther north, near Vail Pass and Grand Mesa. Patchy lynx reproduction continued, with no reproduction in 2007 and 2008.

In 2010, Shenk concluded that if the patterns of annual reproduction and survival that had occurred since the reintroduction were to continue over the next twenty years, this lynx population would sustain itself for the long-term. However, since then, growing scientific knowledge about climate change and its effects on snowshoe hares has created uncertainty about the long-term viability of the Colorado lynx population. On the positive side, recent large dispersals by Colorado lynx into places like New Mexico indicate that at least they'll maintain their genetic diversity.[21]

Lynx Habitat Needs and Climate Change

On multiple levels, climate change is having significant effects on lynx conservation. On the most fundamental level, climate affects lynx because it affects its snowshoe hare prey. As the climate continues to warm, boreal regions in

Canada and the United States, as well as the southern range of lynx habitat, will have progressively less snow cover. We've seen that snowshoe hares turn white in winter as camouflage against snow and brown in summer to blend into the forest duff. This snowshoe hare survival mechanism to avoid predation developed over millennia. Hormones, triggered by a signal from the pineal gland, cause this color shift. The pineal gland gives its signal in response to lengthening spring days or shortening fall days. This means that day length drives these color changes—not snow cover or temperature.

Can snowshoe hares adapt to the decreasing snowpack resulting from climate change? To answer that question, Scott Mills radio-collared 250 snowshoe hares in Montana and Washington. After three years, he found that these hares hadn't made changes in their molting pattern to stay better camouflaged.[22] Indeed, during a recent June field survey at a subalpine lake in Waterton Lakes National Park, Alberta, I found a still-white snowshoe hare standing out starkly against the snow-free forest ground. The hare, easy pickings for any predator, stood frozen like a statue, behaving as if it were still camouflaged against the snow. This color "mismatch" as Mills calls it, is making late spring and early fall particularly deadly times for snowshoe hares. According to wildlife ecologist Laura Prugh, a University of Alaska Fairbanks expert on mesopredator effects in food webs, the mismatch is creating easy hunting for lynx in fall and spring—a short-term benefit. However, its long-term effect will be snowshoe hare decline.[23]

Ultimately, natural selection (survival of the organisms with genetic traits that enable them to survive and reproduce) could be the key to hare persistence. Mills and his colleagues point out that hares don't molt in some places in the world, and molting is an inherited genetic trait. If molting has a negative effect on hare survival in a warming world, then natural selection should favor hares that don't molt. Nobody knows whether natural selection can happen quickly enough to save snowshoe hares from extinction.[24]

In Denali National Park, Prugh and her doctoral student Kelly Sivy are examining the ecological role and population dynamics of coyotes and other mesopredators such as lynx. Specifically, they're evaluating the influence of wolves, prey availability, and snowpack characteristics related to climate change on Denali's coyote population. They're also examining whether increased coyote presence in this park affects the abundance of other mesopredators, such as lynx and red foxes (*Vulpes vulpes*), with which coyotes compete.

An intriguing aspect of their investigation involves looking at the hunting efficiency of two species that prey on snowshoe hares: coyotes and lynx. Coyotes have very small feet and can be much heavier than lynx (in Alaska, coyotes weigh 35–40 pounds and lynx 20–25 pounds). This means that in deep, soft snow, lynx can move more easily and swiftly with their big feet than coyotes, which will sink. Therefore, deep snow gives lynx a hunting advantage over coyotes and enables them to capitalize on available snowshoe hares. Because of climate change, we're seeing more frequent rain-on-snow events: episodes of heavy snow followed by warm temperatures that lead to rain. After the rain, the temperature typically drops again, creating an ice crust on the snow. This hard surface gives coyotes a hunting advantage. Thus, climate change may be making hares more vulnerable to coyote predation.[25]

Some believe that lynx are living on borrowed time. Managers point out that because ecosystems are so dynamic and complex, anticipating the effects of climate change on lynx and creating management strategies to address issues such as habitat loss will be a formidable task. It'll take more science and working within an adaptive management framework (i.e., learning by doing), to find the best ways to protect habitat. Part of the solution may lie in identifying key lynx-movement corridors and creating plans that will help keep those corridors open, enabling lynx to disperse. Such corridors will help maintain genetic diversity among lynx populations, thereby increasing lynx ecological resilience.[26] Another part of the solution lies in taking a closer look at how lynx fit into whole ecosystems and carnivore food webs, via trophic cascades relationships.

The Ecological Effects of Lynx

Like most carnivores, lynx have strong effects in ecosystems. Beyond the lynx–snowshoe hare food web relationship, researchers have suggested other interesting trophic links. In the early 2000s, wolves began to drift down from Canada and across the border from Idaho into Washington State. This natural wolf recolonization inspired ecologists such as William Ripple and his colleagues to conduct an exercise in scientific thinking to consider potential impacts that an apex predator like the wolf would have on the intricate workings of the lynx–snowshoe hare food web. At the time, Washington had low hare and lynx populations and a high coyote population. Coyotes had recently expanded their range and abundance there. Wolves prey on coyotes. What if the wolf's return to the Pacific Northwest could indirectly improve lynx conservation, via trophic cascade effects?

Ripple and colleagues hypothesized that two mechanisms would drive wolf–coyote–lynx–snowshoe hare trophic cascades. First, by killing coyotes, wolves would reverse the *mesopredator release* that had occurred when wolves had been extirpated from this region nearly a century earlier. When humans hunted wolves to extinction, they removed an important check on coyote numbers—creating a "release" on the numbers of this mid-sized predator. Coyote numbers increased, which put more pressure on lynx via competition for food resources such as snowshoe hares. Therefore, by reducing coyotes in this system, a returning wolf population could indirectly create ecological benefits for lynx. The researchers further hypothesized that because coyotes also prey on lynx, a reduction of coyote numbers by wolves would release predation pressure on lynx.[27]

Second, in western North America, south of the US-Canada border, wolves prey primarily on elk and deer. When elk and deer numbers are high, these herbivores can suppress shrubs via heavy browsing. Ripple and colleagues hypothesized that the wolf's return would reduce elk and deer numbers and also change their behavior, as has been found in places like Yellowstone by John Laundré and others.[28] Elk and deer need to stay alert in order to survive in areas where wolves exist. This means keeping their heads up and spending less time standing in one spot with their heads down, as they typically do when there are no wolves in a system. Via this predation risk mechanism, wolves could indirectly reduce browsing pressure on shrubs, which would improve snowshoe hare habitat, thereby benefiting lynx. While these scientists didn't collect data to measure these relationships and test their hypotheses, they suggested that it would be very feasible to do so.[29]

John Squires and his colleagues responded that while these hypotheses may sound plausible, empirical evidence doesn't support them. Wolf–coyote interactions are far from simple. In the early years of the Yellowstone wolf reintroduction (1995–2000), wolves reduced coyote numbers. However, these effects didn't continue, as coyotes adapted to living with wolves and rebounded in number. And in some systems, positive relationships between coyotes and wolves occur, whereby coyotes benefit from scavenging wolf kills. Additionally, the scientific community hasn't found much evidence of coyote predation on lynx in the contiguous United States. For example, a ten-year study in western Montana failed to find any instances of coyote predation on lynx. Further, the winter diet of coyotes sympatric with southern lynx consists primarily of carrion, with snowshoe hares a minor component of coyote diet. Thus in the southern portion of lynx range, little evidence exists of competition between lynx and coyotes for snowshoe hares.[30]

Squires and his colleagues suggested that the second hypothesis, in which wolves indirectly change elk and deer densities via predation, and elk and deer foraging behavior via predation risk, may not hold up either. Scientists have found great variation in these relationships, based on context.[31] We looked at some of the many different ways context can affect food web relationships in chapter 2 of this book.

Geographical context may be a key influence on the relationships described above. Although Prugh and others have found no evidence of coyote predation on lynx in Alaska, snowshoe hares are coyotes' primary prey there. Evidence of coyote predation on snowshoe hares in Alaska suggests that some of the patterns hypothesized by Ripple and colleagues may hold better in the far north than in places like Washington. Prugh and Sivy's research will shed further light on these interactions.[32]

Thinking on a broader food web scale and considering trophic and competitive interactions among species (e.g., among wolves, coyotes, and lynx) are important and often overlooked aspects of conserving threatened species.[33] Ideas on how apex predators like the wolf can affect lynx interactions with whole food webs have significant merit, but such ideas still have to be tested empirically.

Coexisting with Lynx

Few conflicts between humans and lynx have been recorded. Lynx don't kill livestock. They don't threaten humans. Actually, the strongest point of conflict between lynx and humans has to do with forest management.

Since USFWS put the lynx on the list of threatened and endangered plants and animals with a designation of "threatened" in 2000, the agency is still in the process of refining its definition of lynx critical habitat. The 2009 draft lynx critical-habitat document describes it as mature forest with complex structure. This precludes intensive timber harvest in areas inhabited by lynx or identified as critical habitat.

Beyond the natural-resources-use issues identified in this chapter, climate change presents a much more difficult conservation challenge for lynx. It's adding further complexity to predator–prey relationships that are already among the most complex in the natural world. Some of the research being done on how climate change may be shifting lynx food web relationships may help us find ways to create a more sustainable lynx population.[34]

In the coming years, I'll be watching to see when snowshoe hares and lynx return to the regrown forest around Moosewillow. I look forward to the time when our winter ski forays once again yield stories of lynx lying in ambush for foraging snowshoe hares. Time will tell how snowshoe hare and lynx populations respond to forest management and climate change.

Each carnivore has an Achilles heel. Other members of the cat family aren't as vulnerable to climate change as lynx, nor are they as tied to a single prey species. In the next chapter, we'll look at another felid which, while far more adaptable than lynx, faces substantial conservation challenges in the human–wildland interface.

Cougar (*Puma concolor*). (Drawing by Lael Gray.)

CHAPTER EIGHT

Cougar (*Puma concolor*)

Years ago, before I went to graduate school, I was a naturalist specializing in songbirds. I spent one summer doing bird surveys near my northwest Montana home for the Cornell Lab of Ornithology. My task back then was simply to determine whether small habitat patches could sustain nesting bird populations. I deepened my knowledge of birds that summer all right, but in the process learned more than I had bargained for.

For the Cornell study, I began by identifying intact forest patches within a matrix of more-developed lands. My study sites ranged from an 80-acre state park located next to a small town to a 200-acre old-growth forest next to a recent clearcut. In these patches, during breeding season (late May through early July in the northern Rockies), I counted the songbirds present and whether certain key species, such as yellow warblers, had nested. To do this, I conducted point counts, which involved recording all the birds detected visually or by song during a 45-minute observation period.

I happily settled into a routine of visits to my stations, relishing this still morning time. Working alone, I'd enter a forest patch with a clipboard, binoculars, and bird books in hand. Upon arriving at each station, I'd arrange my gear

191

on the ground, where I could reach things easily, and sit cross-legged on the forest duff, senses wide open.

One early June morning, I was sitting in the state park in an old-growth Douglas fir (*Pseudotsuga menziesii*) grove. The forest was especially still that morning. Sunlight shafted down through openings in the high canopy overhead. Wild orchids sprouted in the filtered light between the widely spaced ancient trees. The syncopated beats of a northern saw-whet owl's (*Aegolius acadicus*) mating call echoed through the forest nave, but the songbirds were quiet.

Twenty-five minutes into my observation period, a skunk (*Mephitis* spp.) sprayed musk about a hundred feet behind where I sat. It took a few seconds for the incongruence of this event to register in my brain. Then the hairs on the back of my neck stood up, and I had an irresistible urge to get up and leave. I felt terrified, but I didn't know why. I scooped up my field gear and without hesitating walked briskly out of the forest on a narrow trail. When I reached the parking lot, I got into my car and drove off, feeling like an idiot. I'd never been spooked in the woods and couldn't understand why I felt this way.

Later that day, the park ranger on duty informed me that, about a minute after I had left, a cougar (*Puma concolor*) had emerged from the forest on the same trail that I'd been on. It was a large animal who'd been stalking and attempting to attack humans all week. A few days later, wildlife managers captured him and found him sick and emaciated, his teeth badly broken and worn. They euthanized him, the safest and most humane thing to do. That skunk had sprayed because the cougar had been approaching me from behind. While I'll never know whether the cougar would have actually attacked me, the skunk and my own survival instincts had kept me safe.

This experience surprised me, because for years our family had been coexisting peacefully with the cougars on our land. At night, we'd occasionally hear the unearthly caterwauling of females in heat. Often cougars would kill deer within a stone's throw of our cabin. We'd come upon these kill sites, characterized by piles of shorn deer hair, because cougars strongly dislike biting into hair. Cougars use their razor-sharp front teeth to shear off their dead prey's hair before feeding. Like most children, our daughters spent a lot of time playing outdoors. Over the years, not once had we felt threatened by these predators.

Several years after my cougar experience in the state park, I began my doctoral research on trophic cascades and wolves (*Canis lupus*) in the northern Rockies. For this research I was studying whether wolves and other large carnivores indirectly affected how aspens (*Populus tremuloides*) grow, by both killing and scaring prey. Researchers had found that wolves make elk (*Cervus elaphus*) more wary, and that this causes elk to consume woody species, such as aspens, more sparingly. This enables aspens to grow and creates habitat for birds.[1] To see if these relationships held in my study areas, I was measuring bird diversity.

Using birds to take the pulse of an ecosystem, so to speak, isn't new. Since the 1800s, naturalists and ecologists such as Charles Audubon, Joseph Grinnell, and Aldo Leopold have measured the diversity of bird species as a means of learning about the ecological health of a system. The reasoning behind this is that a place with a rich bird population must contain sufficiently large patches of habitat. Further, Leopold found that in order to support an abundant and rich bird population, these patches must contain both a flourishing plant community and active predation.[2]

During the second year of my study, I worked in Waterton Lakes National Park, Alberta, in a prairie dotted by aspen patches—a habitat type called aspen parkland. I'd been measuring the trees in a male cougar's territory. This large tom had become known to managers for his ability to consistently take down adult elk by himself—a feat few cougars can accomplish. Park wardens had observed him take down a bull elk in broad daylight, so his methods were well documented. He had it down to an art. He'd ambush a bull elk, leap onto its back, break its neck using the elk's antlers for leverage, and then feast for days on each kill. As I worked in that parkland with my crew, we'd come across the cougar's impressive trail of carcasses—one elk every two weeks or so. His kills were likely scavenged by grizzlies (*Ursus arctos*) and wolves, but we had a feeling that he managed to hold his own against the other carnivores.

The cougar let us know that he was aware of us. Some days we'd go in on a trail, measure aspens for an hour or two, exit on the same trail, and find a huge, fresh cougar scat liberally sprayed with urine on the tracks we'd made on our way in. That tom was both curious about us and was marking his turf. But I sensed that he wasn't a threat, focused as he was on elk.

In early July, I'd nearly finished collecting point-count data. I had 40 stations in Waterton, which I surveyed twice during songbird breeding season. Each station lay at the base of an aspen tree, permanently marked with a piece of rebar set into the ground and a numbered aluminum disk affixed midway up the tree's bole. By the middle of the last morning that I could do these observations, I had one more station to survey. This particular point lay at the tip of an aspen stand on the edge of a flat, open expanse of prairie, near a popular picnic area.

I walked toward that point on a heavily used elk trail, my field crew of three in single file behind me. A bunched-up herd of about 200 cow elk, many with new calves at their side, stood next to the trail. Something was off, because the elk didn't take off when they saw us, as they typically did. They seemed very tense. The cows ran around in tight circles, their calves bleating plaintively. As we proceeded, we spotted a cougar scat on the trail, so fresh that it glistened in the morning light. That scat sharpened my senses.

The Canadian prairie is normally a wind-swept place, but not even a mild breeze stirred this morning. Heavy, gunmetal clouds hung on the horizon and I heard the low rumble of distant thunder. Eight in the morning, and already the temperature had crept into the mid-seventies, the air thick with humidity. A large clump of wolfwillow (*Eleagnus commutata*), a fragrant shrub favored by elk for food, grew next to the trail, about twenty feet from the point-count station. When we were about eight feet from the wolfwillow, I saw its silvery foliage swaying vigorously, as if buffeted by a brisk wind. And then I saw what was making the branches move—a cougar's twitching tail protruding from the base of the shrub. Hunkered belly-down under the wolfwillow, the cougar was watching the elk. Cougars often prey on elk calves, so this cougar's intentions were obvious. From tail-tip to where I judged its nose to be, it looked like a large animal. Perhaps it was our friend from the prairie. Surprisingly, my crew hadn't become aware of the cougar.

We were so close to the cougar that if we made any sudden moves, things could go very badly. One of my field technicians, an excitable young man, had a habit of leaping and running away from things—the sort of behavior that could provoke an attack. I didn't feel any aggression coming from the cougar, so I opted to say nothing.

We proceeded to the stout aspen that bore my aluminum disk. Normally, we conducted point counts while sitting on the ground. This time, I asked my crew to do the count standing beneath the tree's leafy canopy, each of us facing a different cardinal direction, ostensibly so that we could keep a closer watch on the strangely behaving elk. I was using a fifteen-minute observation period for this study, which made this count more doable under the circumstances.

Our count progressed uneventfully, with me counting birds and the cougar companionably watching elk. The songbird harmonics this morning included the clarion song of a northern waterthrush (*Parkesia noveboracensis*), the burred trill of the MacGillivray's warbler (*Oporornis tolmiei*), and the strident "sweet-sweet-sweet, I'm so sweet" of the yellow warbler (*Dendroica petechia*). Above their resonant voices, we heard the breathy, hawklike song of the Western wood pewee (*Contopus sordidulus*), punctuated by the repetitive squeaky-toy sounds of the least flycatcher (*Empidonax minimus*). A Wilson's warbler (*Wilsonia pusilla*) streaked by, the bold black cap on its bright yellow body making it easy to identify. Focused intently on the birds, my field crew never noticed the cougar.

At the end of the count, we left peacefully the way we'd come in. The cougar was still in the wolfwillow as we walked by, tail twitching, watching the elk. When we returned to my field vehicle a few minutes later, I debriefed the crew. They were very glad I hadn't told them about the cougar and that we'd completed our point counts for the season.

Given the animal's close proximity to a picnic area, we reported the incident to the Waterton Warden Office right away. Rob Watt, on duty that morning, said, "Would that have been at 8:30 a.m.?"

"How did you know?" I asked.

He chuckled. "Because at 8:35, we observed a herd of elk and their calves bolt across the prairie, as if something was chasing them. We thought there might've been a predator involved."

The cougar had made his move as soon as we left.

The danger cougars present to humans has been overblown. Some of our fears about these powerful predators come from their stealth. Given the thousands of days I've spent in the field over the years, I've only been threatened by a cougar once, during that incident in the old-growth forest patch. The aggressor was an elderly, emaciated individual who in desperation had attempted to turn from wild prey to humans for food. While cougars often come within close proximity of humans, we're seldom aware of them. Given their elusiveness, we still have much to learn about cougars, their role in ecosystems, and how to better coexist with them.

What's in a Name: Cat of One Color

The cougar (*Puma concolor*) is a large, New World member of Felidae, the cat family. It has the largest historic range of any land mammal species in North and South America: from the southern Yukon to Tierra del Fuego, and from the Pacific to the Atlantic. Its adaptability and efficiency as a predator helped it survive the Pleistocene extinctions of other North American felids, such as the saber-toothed cat (*Smilodon*). Ten to twenty million years ago, the felids evolved into several cat genera. Originally described as *Felis concolor* ("cat of one color") by Dutch taxonomist Carl Linnaeus in 1758, in the 1900s the cougar became known as *Felis (Puma) concolor*, with *Puma* a sub-genus of *Felis*. In 1973, taxonomists reclassified *Puma* as a separate genus.[3]

As they did with wolves, taxonomists had their way with cougars, splitting them into myriad subspecies and then lumping them back together, based on evolving scientific understanding of speciation. Until the 1970s, taxonomists believed that 32 cougar subspecies existed, based on skull and skeletal measurements. Then in the 1990s, genetic analysis determined that only five cougar subspecies exist, with two in North America: *P. c. cougar* (ranges in much of North America) and *P. c. corryi* (limited to Florida).[4]

Puma concolor has more common names than just about any other wild animal. Popular English names for it include mountain lion, panther, and catamount (which means "cat of the mountains"). The name *cougar* is a combination of the two Brazilian words for the jaguar (*Pantera onca*): *cuacuara* and *guacuara*. The name *puma*, adopted in the late 1700s by the Spanish conquistadores, comes from Peruvian Quechua Indians, and means "powerful animal."[5] By whatever name we call it, this apex predator has stirred our imaginations across the millennia and made the hearts of both prey and humans beat a bit faster.

Cougar Natural History

One of the largest members of the cat family in the Western Hemisphere, the cougar is a wide-ranging, cryptic carnivore. This generalist species tends to blend into the background—no small feat considering that males weigh between 110 and 232 pounds, have a total length of up to nine and a half feet from nose to tail-tip, and a shoulder height of up to 31 inches. Females are about 40 percent smaller than males.[6]

This predator's physical traits have evolved for stealth and making rapid kills. Cougars have tawny coats, which can vary among individuals from nearly white to russet. This coloring enables them to blend easily into forest duff, rock outcroppings, and desert landscapes. Their powerful shoulders and front legs are made for ambushing. They grasp their prey with scimitar-like retractable claws and swiftly deliver killing bites with sharp canine and carnassial (shearing molar) teeth.[7]

Cougars have a relatively low reproductive rate. Males can reproduce at the age of three, and females at two. The lag in male sexual maturity helps prevent inbreeding among siblings. Breeding can occur year-round. Females den in caves on high ledges, cliff walls, or even in relatively inaccessible vegetation thickets. After a three-month gestation, they give birth to one to six cubs, with an average litter size of three. The female has sole parenting duties. With the exception of females with cubs, cougars are solitary. As is the case with bears (*Ursus* spp.), adult males sometimes kill and eat cubs.[8]

Like many large-mammal young, cougar cubs mature slowly. The blue-eyed, spotted cubs stay with their mothers for up to two years. However, by the time cubs are seven to twelve months old, their mothers spend most of their time ranging more than 220 yards away from them. Adolescent cougars have pale coats with dark spots on their flanks. Their spots fade when they reach adulthood at two and a half years of age.[9]

Cougars need room to roam, males more so than females. Males have a home range of 75–150 square miles, with female home ranges one-third of that. However, range size depends on many factors, including prey availability and habitat quality. Both sexes mark their territories with scrapes—parallel digging marks made with their hind feet. Cougar scrapes can contain urine, but usually don't contain feces. They heap their scrapes with dirt and leaves, to create obvious territorial markers. With radio collars on cougars, scientists are learning that this species may have more complex and flexible social dynamics than previously thought. For example, in the California Sierra, groups of cougars overlapped their ranges to obtain better access to migratory deer (*Odocoileus* spp.) for food.[10]

Like other large carnivore species, dispersal varies by gender for cougars. Males make large dispersals to find mates and avoid other adult males. For example, in the 1990s a young male traveled from his birthplace in northern Wyoming to Colorado, near Denver, covering a distance of 300 miles in the process. And in New Mexico, young male cougars have dispersed across mountain ranges, an interstate highway, and open desert, taking several months to make these daunting journeys. Females tend to settle in or near their mother's home range.[11]

Cougars are obligate carnivores, which means they can only digest meat. They mostly eat deer, but they also consume bighorn sheep (*Ovis canadensis*), elk, moose (*Alces alces*), and a variety of small mammals. Porcupines (*Erethizon dorsatum*) turn up surprisingly often in their diet. Because of their high energy needs, breeding females have difficulty surviving in areas without large ungulate prey.[12]

Cougars need surprise for successful hunting. Built for agility and brief bursts of speed, they're unable to sustain long-distance, high-speed chases on open terrain, because their cardiovascular system can't handle such activity.

Cougars often kill by jumping on their prey's back and severing the spinal cord. Through the use of such ambush tactics they can kill animals several times their size.

In addition to shorn hair, cougars leave other tell-tale signatures at their kill sites. I came upon one of these signatures while putting in track transects for my research in Glacier National Park. A transect is an imaginary line in a landscape that ecologists use to gather data. We typically pull a surveyor's tape to mark our sampling line, collect data along that line, and then pull up the tape, leaving no trace that we've been there. I was documenting use of the landscape by both large carnivores and their prey along transects by counting animal signs (e.g., tracks, scats, carcasses). My transects ran along predetermined compass bearings and were six-tenths of a mile long. My daughter Bianca, who was helping me in the field on this late spring morning, was nearly done pulling the tape that marked the first transect of the day. She was moving toward a serviceberry (*Amelanchier alnifolia*) and aspen thicket, when she stopped in her tracks and wrinkled her nose.

"I smell a carcass," she said.

The unmistakable heavy, metallic scent of fresh blood wafted from the thicket. I looked around carefully and didn't seen any predators. We were nearly done with this transect, so we finished it, with me walking alongside Bianca as she resumed pulling the tape into the thicket. Inside the thicket, we found an elaborate mound of dirt, twigs, and leaves three by four feet in size. A hoofed leg protruded from the pile: a freshly killed deer. Drag marks extended into the thicket from a nearby grassy opening, where the deer had been killed and then moved twenty feet into the thicket. Plucked deer hair lay scattered about.

"Cougar," Bianca said softly. She was an expert at this, having come upon many such caches while playing in the woods where we live.

I nodded and placed tick marks in my data notebook under the cougar and carcass columns.

To our human eyes, this cougar-built monument looked particularly well-crafted and aesthetically pleasing. The cougar had heaped progressively finer layers of earthen material on the carcass, and then topped the whole thing with alternating layers of aspen and serviceberry leaves. We lingered only long enough to photograph the site and collect the coordinates.

After killing an animal, cougars eat their fill and then cache the leftovers for future consumption. They return to cached carcasses and feed on them for several days. In this case, we reported the cougar kill to the park law-enforcement ranger. The cache lay about 50 yards from a picnic area, which the ranger promptly closed. We kept an eye on the cache over the next two weeks. No other large carnivores disturbed it. The cougar returned daily to feed, reducing the carcass to little more than bones within a week.[13]

Predation rates and the ability of cougars to control prey numbers vary. On average, adult male cougars kill one deer every eight days. However, some studies have found much lower predation rates. For example, John Laundré found male cougar summer kill rates in Idaho and Utah of one deer every fifteen days. In the southwestern United States, Logan and Sweanor found that while deer die of many other causes in addition to cougar predation, these other causes don't effectively keep the deer population low. However, in Idaho, Laundré found that winter snow depth had a more significant impact than cougar predation on the decline of a mule deer (*O. hemionus*) population.[14] These findings may differ because of differences in geography and climate. Cougar predation has threatened the restoration of bighorn sheep populations, and in Nevada, after a decline in mule deer, cougars nearly wiped out a local porcupine population.[15]

Predation by cougars (or any other predator) can remove surplus prey populations. From Alberta to Colorado, researchers have found that in any given year, cougars remove anywhere from 5 to 20 percent of the deer population. By removing already doomed animals from a population, cougar predation compensates for other forms of mortality, such as starvation or disease. However, if a prey population is fit and in good health overall, then cougar predation will remove additional individuals to those that would die of other natural causes. Where deer are struggling to survive, cougar predation helps reduce their numbers, so that those that remain have a better chance of living.[16] Conversely, cougar numbers are closely linked to prey numbers. As prey density increases, more cougars can survive and reproduce. As prey density declines, cougar number decline.

Where more than one species of large carnivore shares an area, competition between these species can influence cougar relationships with their prey. For example, in Glacier National Park, which had been recolonized by wolves, Kyran Kunkel found differences in prey selection by cougars and wolves. While white-tailed deer (*O. virginianus*) composed the greatest proportion of both wolf and cougar kills, elk and moose composed a larger proportion of wolf kills than cougar kills. However, also in Glacier, I found a cougar-killed white-tailed deer ten yards in front of a wolf den. The cougar had actually killed the deer in spring *while* the wolves were at the den (per my radio-collar data). And on the High Lonesome Ranch in Colorado, which has substantial cougar and black bear (*U. americanus*) populations, Mark Elbroch found bears frequently scavenging cougar caches. Because bears were stealing so many of the cougar kills there, cougars may have had to kill more deer to survive. These examples illustrate the complexity of how cougars interact with other large carnivores in areas where they occur together.[17]

Cougars can prey on pets and livestock. Domestic sheep provide the most common cougar target. Cougars have been known to kill five to ten sheep at a site, eat only one or two of the carcasses and then leave the rest. They also kill cattle calves. The highest losses occur in sheep summer range and in spring during calving season. These losses can be prevented by using guard dogs or by having cows calve later in the year. Cougar depredation on pets is uncommon and occurs primarily in the urban-wildland interface, where new housing is encroaching on high-quality cougar habitat.[18] For example, a friend built a house in San Luis Obispo County, California, in a new housing development set against the unpopulated oak (*Quercus* spp.) hills. One morning, she opened her kitchen French doors and let her tiny Lhasa Apso dog out. Less than a minute later, a cougar streaked like lightning across her yard, picked up the dog in its mouth, and left. She never saw her dog again. Her young children, who were in the kitchen eating breakfast at the time, witnessed the whole traumatic incident.

Cougars occupy a broad range of habitat types. Their primary habitat needs are abundant hiding cover and ungulate prey. They use riparian areas for cover, as well as woodlands and rock ledges. Rarely found on valley bottoms, cougars prefer steep, mountainous country and coniferous forests. However, they can thrive in just about any kind of habitat that provides hiding opportunities, such as the rainforests of the Pacific Northwest, the high desert of the southwestern United States, and wetlands. Using radio-collar data, scientists have confirmed that cougars use *ecotones* (the place where two habitats come together, which forms an edge, such as between forest and prairie) for stalking.[19]

Although primarily a wildland species, cougars sometimes inhabit areas with high human presence. In Southern California, Paul Beier and his colleagues have found that cougars in the urban-wildland interface spend most of their time on the wildland side. However, in Washington, cougars had similar space-use patterns in both wildland and residential areas, and in selected forest patches, reserves, and forested corridors. And in Mexico's Chihuahuan Desert, Laundré and his colleagues found that cougars spent more time in areas with abundant mule deer and low road and human density. Because of the potential for conflict between humans and cougars, low human density represents a third cougar habitat need.[20]

Cougar Conservation History

By the 1960s, unregulated hunting by humans and loss of habitat had greatly reduced cougar distribution, to the point that this species had gone extinct in large areas of its historic North American range. As a result, cougars were confined to the western mountains and the southern Yukon. No reliable estimate exists for the population before the 1970s.

Since then, cougar numbers have been increasing. In the 1990s, Nowell and Jackson estimated that 10,000 cougars roamed western North America. By 2010, this number had become much larger; 5,000 cougars occurred in Oregon alone, with a similar population in California. However, due to this species' elusiveness, accurate cougar counts don't exist for North America. Figure 8.1 de-

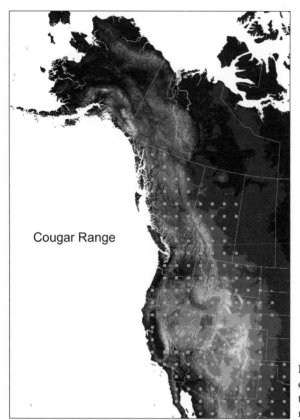

Cougar Range

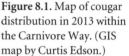

Figure 8.1. Map of cougar distribution in 2013 within the Carnivore Way. (GIS map by Curtis Edson.)

picts cougar distribution within the Carnivore Way. Note that gaps exist in our knowledge of cougar habitat in Mexico as well as the eastern edges of its North American range, due to lack of radio-collar data in those places.[21]

Since 1990, cougars have been dispersing into the Midwest and the northeastern United States. They've recolonized North and South Dakota and the Pine Ridge country of northwestern Nebraska, expanding their range eastward. Numerous verified sightings have occurred in Wisconsin and other midwestern states. And in June 2011, a non-radio-collared young male cougar turned up in Connecticut, road-killed by a sports utility vehicle. The 140-pound cougar was in good health at the time of his death. DNA tests furnished genetic evidence that he'd traveled 1,500 miles from the Black Hills of South Dakota, via

Wisconsin and Minnesota, eastward to suburban Connecticut. Multiple cougar sightings in Pennsylvania, one of the states through which he'd passed, provided further evidence of his continental-scale journey. He set a record for the longest dispersal ever recorded for his species and provided the first confirmed cougar presence in Connecticut in over 100 years. Cougars have been turning up in other surprising places, sometimes with tragic outcomes. For example, in April 2008, police shot and killed a cougar in Chicago, Illinois.[22] These dispersals have inspired expert wildlife tracker Sue Morse to develop noninvasive monitoring strategies for cougars. Morse and cougar biologist Harley Shaw have teamed up to educate state-agency managers and the public on how to identify cougar tracks and sign in order to help document dispersals.[23]

Cougar biologist Maurice Hornocker believes there now may be more cougars present in the United States than before the European settlement of North America. This increase may be due to the fact that humans manage deer and elk numbers at far higher levels than may have occurred before European settlement, and also due to ranching, which provides additional potential prey (e.g., sheep, calves) for cougars. However, today cougars face new threats, such as interstate highways with four lanes of heavy traffic.[24]

Until the mid-1960s, cougars received no protection throughout their North American range. State agencies paid bounties to hunters and trappers as incentives to reduce cougar numbers. These bounties ranged from $20 to $30 in the 1920s and from $50 to $60 per cougar by the 1940s. A skilled trapper could clear up to $20,000 annually in the 1920s—a small fortune back then. But by the mid-1970s, due to low carnivore numbers and changing conservation values, Canadian, US, and Mexican provinces and states had eliminated bounties.[25]

Since the mid-1970s, cougars have been managed through sport hunting, considered by wildlife agencies to be a form of protection because regulated cougar-hunting quotas control the number of cougars killed. In all western states except California, cougars can be hunted legally. Methods range from archery and rifle seasons to hunting with hounds. This latter method has become one of the most popular forms of cougar hunting, because of its effectiveness. A team of specially trained dogs locate a cougar's scent, find the animal, and pursue it until it climbs a tree. When treed, the cougar can then be shot easily by the hunter. Animal-rights groups have protested hunting with hounds as inhumane, citing that it gives humans an unfair advantage. Scientists also use hounds to capture cougars for radio-collaring.

In Canada and Mexico, cougars are freely hunted. However, outside of Mexico, most Latin American countries completely protect cougars and don't allow hunting them at all. The International Union for the Conservation of Nature (IUCN) considers the cougar a species of "least concern." Nevertheless, the organization acknowledges that this species may be declining in the Western Hemisphere, although the scientific literature contains some ambiguities about this species' conservation status. Some sources report population increases and others report population decreases due to anthropogenic causes. Problem areas for this species lie in Latin America, due to poaching and habitat fragmentation. In North America, it is expanding its range. All of this suggests a cautious approach to cougar management. [26]

Hunting by humans may threaten cougar persistence in a variety of ways. Cougar killing that exceeds 40 percent of the population for four consecutive years can have significant negative impacts on cougar demographics. Additionally, human hunters tend to kill cougars as trophy animals, targeting large adult males. Hence, hunting eliminates the oldest, most stable members of a cougar population. This leaves a void, enabling young, inexperienced males to disperse into the vacated territories. Young males are more liable to get into trouble by depredating on pets and livestock. Further, hunting may not always achieve managers' objectives of reducing the cougar population. In Washington, hunting cougars had little impact on cougar numbers, because new cougars quickly dispersed into areas where cougars have been killed by hunting.[27]

Habitat fragmentation represents the single largest threat to cougar conservation. Cougar natural history requires male dispersal in order to maintain genetic diversity and minimize conflicts between male cougars. Two phenomena, the *edge effect* and the *Allee effect*, make cougar dispersal challenging, ineffectual, and sometimes even impossible. The edge effect involves a situation where unsuitable habitat surrounds a local cougar population, making it difficult for individuals to disperse. The Allee effect occurs when males disperse into areas with a low density of other cougars—so low that they fail to find mates. Warder Clyde Allee never formally defined this effect. But according to Philip Stephens and his colleagues, Allee saw it as "a positive relationship between any component of individual fitness and either numbers or density of other animals of the same kind." In other words, a larger group size or some degree of crowding may stimulate reproduction and improve survival in adverse conditions. These dispersals lead to genetically nonfunctional populations.[28]

Given what we know about habitat fragmentation, conservation biologists don't think that cougar conservation is as simple as setting limits for human hunters. Cougar metapopulations have within them local *source* and *sink* populations. A source population occurs in high-quality habitat (e.g., one with plenty of ungulates and hunting cover), in which cougar numbers become robust. Source populations send out dispersing individuals to create new populations. A sink population, on the other hand, is one that lies in poor habitat, or next to high human development, so that dispersing individuals are unable to survive. This causes the sink population to become genetically stressed. A metapopulation approach to managing cougar hunting in the western United States could mean identifying source and sink populations, closing a significant amount of cougar habitat to human hunting in order to ensure long-term population survival, and working within established hunting-management units to make decisions that are more science-based.[29]

The Ecological Effects of Cougars

Like most large carnivores, cougars undeniably have the potential to wield a powerful effect on ecosystems. Food web relationships between carnivores and their prey create ecological checks and balances. For example, cougars prey on mesopredators such as foxes (*Vulpes* spp.) and raccoons (*Procyon lotor*). Mesopredators raid bird nests for eggs, a fast and easy protein source. Thus, by limiting mesopredator numbers, cougar presence can indirectly increase songbird survival. Such relationships represent a trophic cascade in which an apex predator's direct effects on its prey cascade through a food web to affect other species, such as the birds in an aspen stand.[30]

Where there are more deer than elk, cougars may be driving other types of trophic cascades. For example, in Zion National Park, William Ripple and Robert Beschta found a correlation between cougar presence and the growth of riparian vegetation. Here, by reducing mule deer numbers and changing deer behavior (e.g., deer avoid areas of high cougar presence), cougars may be indirectly enabling previously over-browsed riparian vegetation to grow and thus to provide habitat for other species, such as butterflies.[31] In Yosemite National Park, these researchers looked at a similar relationship among cougars, mule deer, and black oaks (Q. velutina). Here, they found a correlation between cougar presence and the growth of oaks, a preferred deer food.[32] However, these cougar-caused top-down effects haven't been tested using radio-collar data and are based on an assumption that cougars avoid areas of high human use, such as campgrounds and picnic areas. However, using radio-collar data, many national parks (e.g., Waterton Lakes National Park) have reported frequent use of such areas by cougars. Much more research is needed to determine how and when cougars drive trophic cascades.

According to Doug Smith, leader of the Yellowstone National Park Wolf Project, and Rolf Peterson, who has conducted wolf research on Isle Royale National Park for nearly five decades, the way in which cougars affect food webs mainly has to do with this species' social habits. Unlike wolves, cougars are solitary stealth hunters that don't form social groups. Because of this, where wolves and cougars occur together, wolf presence has greater impact than cougar presence on top-down trophic interactions.[33]

One of my research areas, the Saint Mary Valley of Glacier National Park, didn't contain a breeding population of wolves during the course of my study. However, it did contain abundant cougars, black bears, and grizzly bears, with ten times as many elk present as deer. There, despite a high number of cougars, the elk were relatively complacent, standing around in one spot browsing the aspens down to ankle height. Apparently, cougars weren't driving a trophic cascade in that system.[34]

Given such findings, perhaps the cougar's greatest universal impacts on food webs occur through the food they inadvertently provide for other species—called *food subsidies*. By killing and caching ungulate carcasses, other carnivores and omnivores, such as bears and ravens (*Corvus corax*), benefit via scavenging. For example, in South America, wildlife ecologists Mark Elbroch and Heiko Wittmer found that cougars indirectly feed a wide range of scavengers, including Andean condors (*Vulture gryphus*).[35]

Paul Beier and Kyran Kunkel recommend using the cougar as a focal species for conservation. They define a *focal species* as one that requires so much of a particular habitat (e.g., one with low human density) that a conservation plan providing enough of this resource will also meet the needs of many other species. Cougars make effective focal species because they require corridors and space. Meeting cougar needs for space will also help conserve other species, such as bears, that also require space. And accommodating cougar habitat requirements will restore ecological relationships triggered by cougars, such as food subsidies to other carnivores and potential trophic cascades.[36]

Are Cougars Dangerous to Humans?

In modern times, a series of fatal cougar attacks on humans has prompted wildlife managers and the general public to take stock of the danger that cougars pose to humans. Beier analyzed cougar-caused human deaths in Canada and the United States between 1890 and 1990. He found nine fatal and 44 non-fatal attacks, and a sharp increase in attacks during the 1980s. Mainly, the victims were children. Juvenile cougars and adult cougars in poor physical condition were responsible for most of the attacks. Sweanor and Logan documented a fivefold increase in fatal cougar attacks on humans between the 1970s and 1990s, attributable to the growing human population in areas that previously had been high-quality cougar habitat.[37]

In 2011, David Mattson and his colleagues analyzed 343 aggressive cougar encounters in the United States and Canada. Only 29 resulted in human fatalities. They found that young cougars in poor physical condition are more likely to challenge humans, and adult cougars less likely to threaten humans, but more likely to cause death. Cougar attacks and subsequent human deaths were more likely if a child was present, and assertive or aggressive behavior by humans (e.g., yelling, throwing rocks and sticks) lessened the likelihood of an attack.[38]

Well-documented examples of coexistence between humans and cougars are uncommon, due to this species' elusiveness. Yet every once in a while a cougar comes along with the makings of legend to provide such an example. In Banff National Park, the 2001 death by cougar attack of Frances Frost while cross-country skiing caused wildlife managers to radio-collar a dozen cougars and study their behavior relative to humans. One of them, a mature male, was chased by hounds up five Douglas fir trees before wildlife biologists were able to successfully dart him and fasten a radio collar around his neck. Formally referred to as Cougar 61 but subsequently dubbed Doug, this 150-pound tom proceeded to capture the public's imagination and teach managers and the public about coexistence (fig. 8.2).

Figure 8.2. Doug the Cougar in Banff National Park, Alberta. (Photo by John E. Marriott.)

Collar data quickly showed Doug's propensity for hunting near human development. Sometimes he'd take deer down in one of the park picnic areas. Managers surmised that these takedowns had probably seemed like a good idea under cover of darkness. Once the sun came up, Doug found himself amid lots of human activity, yet he continued to hunt in the area. Managers and the public grew fond of Doug, who never threatened humans. His collar data showed further that while he was in the park backcountry, which had lower human activity, he became more diurnal. Steve Michel, the Banff human-wildlife-conflict specialist, cites Doug as an example of an animal that has learned behaviors (e.g., hunting at night) that allow him to coexist with humans. Doug eventually died at the ripe old age of fourteen, when he fell through the ice while chasing an elk. He left behind an untarnished reputation.[39]

For those of us who live, work, and recreate in cougar country, it's impossible to avoid cougars entirely. As an ecologist, I work deeply immersed in cougar-filled landscapes. Consequently, I've had a broad spectrum of experiences with this species—everything from being stalked to spending many days afield peacefully working in areas with large populations of cougars. When I'm not working in the field, I run to stay fit. There's no cougar-free place near my home. I run anyway, my senses wide open as I take in great lungfuls of clean mountain air. I don't feel afraid, but neither am I complacent. In a sense, you could say I'm like one of those vigilant elk: wary, but still going about my business.

Perhaps my most astonishing lesson about cougars occurred while I was working in Glacier National Park's North Fork Valley. My field crew and I were staying in a house owned by The Nature Conservancy (TNC), just outside the park's west boundary. TNC allowed researchers whom they perceived were doing work that contributed significantly to conservation to use that house. The locals called it The Palace, because it was far grander than the typical residence in that area. To understand better why they'd given that name to the house, it helps to know a little about the North Fork and the "locals." One of the wildest places in the lower contiguous United States, this rough valley is inhabited by a handful of people sufficiently stalwart to brave the minus-40-degree-Fahrenheit winters and not mind either the lack of electricity and most of the other amenities of civilization, or an ultra-high density of carnivores of all kinds. Most people here live in simple, century-old cabins made of hand-hewn logs. The Palace, an elegantly rustic 4,000-square-foot cedar home entirely powered by solar energy, had been featured in *Architectural Digest* a few years before our stay. It stood in a totally secluded spot, in the center of a large meadow, with a pond next to it. A thick lodgepole pine (*Pinus contorta*) forest surrounded the meadow.

My crew and I had just come in from the field from measuring aspens, having put in a ten-mile day wearing heavy packs. I'd been staying with them at The Palace but had dropped them off to go home and spend a rare night with my family. (I live 100 miles from The Palace, so during this period I usually stayed with my crew and didn't go home at night.) When I came back the next morning and picked up my crew to head into the park for work, I learned about what had happened in my absence.

Hot, tired, and hungry, but feeling good about their work that day, my crew had sat on the house's large, wrap-around porch and unlaced their boots. All at once, a wolf darted across the meadow. That alone wasn't so surprising. While they'd never seen a wolf come so close to the house, there were many wolves in the area. A few seconds later, they heard animal screams coming from just inside the forest edge, about 50 yards from where they sat. It sounded like an animal dying. Thirty seconds later they heard growling and other primal, indescribably wild sounds, branches breaking, the sounds of a battle. Their senses were still reeling from the sound effects when a cougar exploded out of the forest and headed directly toward them at a full run. The animal came within ten feet, so close that they could see that she was a female. Ears flattened, mouth open, lips drawn back to expose her sharp teeth, she snarled but didn't make eye contact with them. They went inside as calmly as possible (which wasn't easy), because rapid, sudden movements could have provoked an attack. The Palace had wrap-around picture windows to go with its expansive porch. This enabled the crew to watch the cougar safely.

Angry was the operative word for that cougar. They'd never seen an angrier animal. She proceeded to circle the house, staying very close, but not looking in their direction. She seemed well-muscled and large for a female, but a bit lean. Welts—what looked like long, parallel clawmarks—were rising along one of her flanks, although her skin was unbroken. My crew perceived no threat coming from her, just intense anger. They photographed her through the windows. She circled the house for ten minutes, and then left. They didn't see or hear any more commotion, so they ate dinner and went to bed.

The next morning, all seemed peaceful when I picked up my crew and we left for the field. Upon returning that afternoon, we went into the woods to check out the scene of the crime. There we found a white-tailed deer carcass surrounded by the dinner-plate sized tracks of a very large grizzly bear sow, and the smaller tracks of her cub. We knew the tracks had been made by grizzlies, not black bears, because of their long claw length. We also saw a spot where there'd been a skirmish—a welter of cougar tracks mixed in with grizzly tracks. The deer carcass had been worked over by the bear, but we could still see its broken neck and torn throat.

Every kill site tells a story. Figuring out that story involves detective work. Putting together all of these clues, we determined that the grizzly, who like all bears at that time of year was hyperphagic and on the prowl for extra calories and protein, had attempted to steal the cougar's kill. A battle had ensued, during which the grizzly bear had gotten the upper hand. The cougar had bolted out of the woods and run toward the perceived safety of humans—a tired field crew unlacing their boots on the porch at the end of a hot autumn day. The savvy cougar knew that a grizzly with a cub wouldn't pursue her so close to humans. She'd continued to circle the house long after my crew had gone inside, until she sensed it would be safe to leave. As for the wolf, we speculated that it had just been passing through, checking out the action in the woods.

This incident helps illustrate the range of cougar relations with humans. Usually cougars exhibit neutral behavior around us, as was the case with Doug the cougar; sometimes they're curious about us, other times the see us as antagonists. On rare occasions they see us as prey. And sometimes they use us for protection from other predators—a behavior that conservation biologist Joel Berger calls *human shielding*.[40]

Coexisting with Cougars without Fear

Restoring cougars to the West brings up questions about balancing the ecological needs of this species and with those of humans. Additional questions raised by the popular media, especially after fatal cougar attacks, indicate that cougar restoration isn't as simple as just wanting them back. As human populations continue to expand and cougar habitat becomes increasingly fragmented, what limits should be placed on human behavior in the urban-wildland interface? What's the role of management? And how do shifting attitudes reflect the demography of human communities? In recent years, two Arizona situations inspired conversations among residents, managers, and scientists about these questions.

In 2001, two hikers on Mount Elden, immediately adjacent to Flagstaff, had a frightening encounter with two cougars. The cats stalked the hikers' dogs and tangled with them. The hikers threw sticks and rocks to chase them off and made their way back to the Mount Elden trailhead and ranger station. The cougars followed the hikers nearly all the way back. Nobody was injured. Many other persons who used Mount Elden for horseback riding also reported cougar encounters and stalkings. Also in Arizona, in Sabino Canyon near Tucson, since 2003, people have reported 100 cougar sightings annually. Cougars sometimes stalk people who use this popular recreation area. To protect humans and cougars, the Arizona Game and Fish Department closes this area periodically.

In 2003, in response to the human-safety issues created by these cases, the US Geological Survey initiated a cougar study that continued for several years. David Mattson, who headed the study, collared several cougars. He and his colleagues found that these cougars were primarily eating elk and deer. Further, these cougars were coexisting relatively peacefully with humans, given the high density of both species in his study area.[41]

Some conservation biologists recommend taking a more specific approach to improve cougar survival and reduce conflicts with humans. Beier suggests maintaining habitat cores that have abundant cover, prey, and low human presence, and an area of at least 400–800 square miles. Corridors linking such cores should have nearly continuous woody cover, underpasses integrated with roadside fencing to prevent vehicle strikes, little or no artificial lighting, and less than one human dwelling per 40 acres. For corridors up to four miles in length, he recommends a minimum width of 400 yards. At the University of California at Santa Cruz, wildlife ecologists Chris Wilmers and Terry Williams, along with field biologist Paul Houghtaling, have pioneered a tracking collar that enables researchers to mark a cougar's location, its rate of travel, and its activity level (e.g., resting or eating). They're using this type of collar to identify cougar travel corridors, particularly those related to dispersal, and to measure cougar energy needs. Based on their findings, Wilmers suggests taking a closer look at how cougars use corridors in response to human development. For example, cougars make greater use of areas that have intermittent human presence. In Banff, the Wildlife Crossings Project has applied corridor-ecology concepts (e.g, identifying cores and corridors) to design a series of crossing structures over and under Highway 1. Cougars are primarily using the underpasses on this project, but they sometimes use overpasses as well.[42]

In addition to such science, David Mattson and Susan Clark's work has shown us that communication among people provides the key to cougar conservation. Management that favors discourse is necessary for coexistence where humans and cougars are liable to come into conflict. Additionally, where a "cores and corridors" approach to cougar conservation isn't fully feasible and reserves are unlikely to be expanded in order to improve cougar habitat, sound management will likely provide the deciding factor for coexistence. Such management needs to be developed in collaboration with a diverse group of people and to incorporate ethics and human values, along with best science.[43]

Cougars are rapidly reclaiming their historic North American range and providing us with the perfect opportunity to learn to live with large carnivores. Next we'll learn about another member of the cat family that shares many traits with cougars, yet is far less flexible in its habitat needs—and consequently is far more at risk.

Jaguar (*Panthera onca*). (Drawing by Lael Gray.)

CHAPTER NINE

Jaguar (*Panthera onca*)

By 1960, the jaguar (*Panthera onca*) seemed to have vanished from the southwestern United States, the northernmost edge of its historic range in the Americas. Wildlife managers dismissed persistent rumors of sightings of this most iconic and mysterious member of the North American cat family as wishful thinking. Then in March 1996, while cougar (*Puma concolor*) hunting in the Peloncillo Mountains on the Arizona–New Mexico border just north of Mexico, rancher Warner Glenn and his daughter Wendy had an encounter that changed everything. On the fourth day of the hunt, Glenn's hounds picked up a wild cat's scent. The hounds trailed the cat in rough terrain through brush-choked rimrock canyons. Cougars usually climb trees when pursued by hounds. This animal didn't do that. After a long chase, the cat ended up on an outcropping, just beyond the hounds' reach. Glenn caught up, tied up his mule, and headed into the fray. To his shock, the cat on the rocks was an adult jaguar. Glenn called off his hounds and photographed the animal before it took off southward, toward Mexico. His photographs were the first taken of a wild jaguar in the United States.[1]

Jack and Anna Mary Childs and Matt Colvin had a similar experience in August 1996, while cougar hunting in the Baboquivari Mountains southwest of Tucson, Arizona. Their hounds treed a cat in a canyon bottom. When the Childs and Colvin approached, the cat at bay near the top of an alligator juniper tree (*Juniperus deppeana*) turned out to be a jaguar. The large animal, which lay sprawled on a limb, one huge paw dangling down, was a male. Childs took out his video camera and proceeded to film the jaguar. At one point the jaguar, having recently fed, lowered his head and napped for a few minutes. They observed him for an hour or so, and then leashed their dogs and departed peacefully.

This particular jaguar, eventually named Macho B by researchers, would have a seminal role in the conservation of his kind. A distinctive mark on his right flank—a broken rosette in the shape of the cartoon character Pinocchio—would make him easily recognizable over time.[2] Five years after this encounter, Childs's haunting images would provide powerful ecological evidence to document a dispersing jaguar population in Arizona. Further, this animal would teach us essential lessons about large carnivore conservation and its challenges. In many ways, this chapter is that jaguar's story.

What's in a Name: Large Spotted Cat

The jaguar is the only member of the genus *Panthera* in the Americas. It occurs in the New World, but descended from Old World felids. Six to ten million years ago the roaring cats, which today consist of the jaguar, lion (*P. leo*), tiger (*P. tigris altaica*), leopard (*P. pardus*), snow leopard (*Uncia uncia*), and clouded leopard (*Neofelis nebulosa*), shared a common ancestor. The genus *Panthera*, which means "large spotted cat," emerged from this ancestor 3.8 million years ago. *Panthera* moved into the New World via the Bering Land Bridge 2 million years ago, along with prey such as mammoths (*Mammuthus* spp.). About 600,000 years ago, the jaguar (*Panthera onca*) separated from the other members of *Panthera*. This split is sufficiently recent in evolutionary time that today jaguars and leopards in captivity can interbreed. Much larger than today's jaguar, the Pleistocene jaguar roamed much of North America and spread into South America. Toward the end of the Pleistocene epoch, between 15,000 and 12,000 years ago, jaguars diminished in size to their current body mass. At the same time, their distribution contracted to what is now the southwestern United States down to the tip of Argentina in South America. During the early Holocene epoch (about 11,000 years ago), their primary American feline competitors became extinct, with the exception of the cougar.[3]

Like other broadly distributed carnivores, the jaguar has many names. Spanish-speaking nations refer to it as *el tigre*. Its English name, *jaguar*, originated from the South American Tupi-Guarani word *yauara*. This indigenous Amazonian word has several meanings, including "wild beast that overcomes its prey at a bound," and "eater of us." For the past century and a half, in the southwestern United States people have been referring to the jaguar as the American or Mexican leopard.[4]

In the 1930s, taxonomists recognized multiple subspecies of jaguar—five in North America alone. They based these distinctions on measurements from a handful of skulls. Unsurprisingly, in the past twenty years, DNA analysis has rendered such distinctions moot. Today, ecologists only distinguish between northern (northern Mexico and southwestern United States) and southern (Central and South American) jaguar populations, since none exist on Mexico's high Central Plateau. However, these populations are genetically linked, so they don't constitute subspecies.

Researchers David Brown and Carlos Lopez Gonzalez use the term *borderland jaguars* for the population within the US states of Arizona and New Mexico and the Mexican state of Sonora. This is the only jaguar population within the Carnivore Way. In this chapter, we'll explore jaguar ecology and conservation status in general, with a primary focus on this northernmost population. Due to habitat differences, these borderland jaguars differ physically and behaviorally from other New World populations of this species.[5]

Jaguar Natural History

The jaguar is the third-largest felid species in the world. While the lion and tiger are larger than the jaguar, in the New World, the jaguar is bigger overall than any other members of the cat family. There is much we still don't know about this species. In fact, some ecologists refer to it as a "ghost species," due to its rarity and the difficulty of detecting it. While researchers are working hard to fill these knowledge gaps, this species' scarcity, particularly in the northern portion of its range (northern Mexico and the southwestern United States), makes such studies challenging.[6]

Across their entire range, jaguars occupy a spectrum of habitat from wooded Sonoran canyons to rainforests to the swampy grassland mosaics of the Brazilian Pantanal to dry deciduous forests. Strongly associated with water, jaguars swim well. The most robust populations have historically occurred in tropical areas with low elevation, thick vegetation cover, and year-round water. However, concealment—closed cover provided by shrubs or trees—represents this species' primary habitat requirement.[7]

Borderland jaguars prefer Madrean evergreen woodland habitat. This pine (*Pinus* spp.) and oak (*Quercus* spp.) subtropical forest type occurs in the Mexican and southwestern US mountains. Sinaloan thornscrub habitat provides transition between Sinaloan desert and Madrean woodland and additional habitat. Thorny deciduous shrubs and trees characterize Sinaloan thornscrub, which often occurs in a mosaic with Madrean woodland.[8]

Two factors influence jaguar body size: habitat and gender. Forest-dwelling male jaguars (e.g., borderland jaguars) average 125 pounds, while those that live in more open terrain (e.g., the Brazilian Pantanal) average 200 pounds.[9] In the 1920s, this size discrepancy inspired naturalist Ernest Thompson Seton to call borderland jaguars "dwarfs." This difference could be due to the fact that forest jaguars live off smaller prey than those in the Pantanal. Additionally, female jaguars tend to be 30 percent smaller than males. Otherwise animals of both sexes are similar in appearance, with stocky, muscular bodies, short, strong legs, and relatively short tails for their body length.

Jaguar vocalizations distinguish them from the rest of the members of the American cat family. The only roaring felid in the Western Hemisphere, jaguars vocalize with a series of cough-like grunts that develop into a full-throated roar. They vocalize to mark their territories, thereby making other animals aware of their presence and advertising their sexual receptivity. In the wild, it can be difficult to determine the location of jaguar sounds, particularly in thick forest.[10]

Jaguars have striking, cryptic coloring, with no two animals exactly alike. They have white fur on their belly and black bars on their chest. Two thirds of their tail is tawny, the last third white with black spots. Charcoal spots in the shape of broken rosettes containing one to three dots mark the rest of their pale-gray to caramel-colored coat. These markings enable them to blend into the forest in dappled sunlight.

Notable differences in coloring exist between northern and southern jaguar populations. Borderland jaguars look faded compared to southern jaguars. While black phase (called *melanistic*) jaguars occur in South America, this color phase has never been observed north of Chiapas, Mexico, and the Central American country of Belize. Melanistic jaguars have charcoal gray fur with faintly visible black rosettes.[11]

Jaguars share natural history traits with other carnivores, but they most closely resemble cougars. I've discussed many of the jaguar traits described below in the species chapters of this book as they pertain to other carnivores. Like cougars, jaguars have a lag between the sexes in reaching sexual maturity, to prevent inbreeding. Jaguar females can reproduce at the age of two and a half to three years, and males at three to four. Breeding occurs year-round. Females den in caves on high ledges and sometimes in abandoned mines and relatively inaccessible vegetation thickets. They have a low reproductive rate. After a 100-day gestation, females whelp one to four cubs, with an average litter size of two. Jaguars resemble other carnivores in their solitary nature (except for females with cubs) and large home ranges. Upon reaching young adulthood, males disperse and establish their own home range. Females make shorter dispersals and sometimes share their mother's home range.[12] Male home ranges can be from 10 to 53 square miles in size, with female home ranges smaller than that. Like cougars, jaguars mark their territories with scrapes—parallel digging marks made with their hind feet. Jaguars also leave five-foot-diameter mounds of leaves and debris.[13]

Jaguars are good hunters, due to their stocky, muscular physique and strong jaws. These nocturnal stealth hunters stalk their prey and attack from the rear, knocking the victim down with a blow to its head, often breaking its neck in the process. With a greater bite force than that of tigers and African lions, jaguars can bite through steel and can easily pierce their prey's skull. With their powerful jaws they can grab prey weighing up to 800 pounds, drag it to cover, and consume it in one or two feeding bouts. Like cougars, jaguars cache prey carcasses, but unlike cougars, jaguars don't bury their caches.

Jaguars prey on large mammals when available. Borderland jaguars primarily subsist on Coues white-tailed deer (*Odocoileus virginianus couesi*) and collared peccary (*Pecari tajacu*). Other foods include coati (*Nasua nasua*), opossum (*Didelphis virginiana*), and other medium-sized mammals. However, jaguars have a broad and adaptable diet, which includes more than 85 species. This variety makes jaguars food generalists. Additionally, jaguars kill proportionately to prey abundance. For example, in areas with more peccaries than deer, jaguars will primarily use peccaries for food. As a food source, particularly during drought years when ungulate numbers are low, and to a greater extent than cougars, Jaguars prey on livestock. In Brazil, livestock sometimes makes up a significant amount of jaguar diet.[14]

Jaguars compete for food with cougars, which also primarily eat deer. While both species occur in the same areas, cougars are far more generalist when it comes to habitat, living anywhere from open desert to high mountains. To avoid conflict, cougars partition themselves spatially away from jaguars (e.g., by using the opposite side of a stream for hunting).

Jaguar Conservation History

From the 1500s to the early 1800s, jaguars lived as far north as what today is the southwestern United States. Spanish explorers documented jaguar depredation on livestock in Mexico as early as the 1600s. Back then, Spain had claimed much of this land and was paying bounties to hunters in order to reduce jaguar numbers. In Mexico, deforestation and the spread of ranching hastened this species' demise, with a sharp jaguar population decline from the 1800s through the mid-1900s. During the US Biological Survey (the precursor to the US Fish and Wildlife Service) predator-removal program from 1918 to 1964, humans killed an unspecified number of borderland jaguars, likely dozens. Due to predator control in all nations, by 2001 the jaguar's known occupied range in the

Americas had decreased to approximately 46 percent of what it was in 1900. Figure 9.1 depicts jaguar distribution in 2013 within the Carnivore Way.[15]

Modern jaguar conservation efforts began in the mid-1960s. Prior to 1966, the Mexican government treated jaguars as game. In the United States, jaguar hunting continued as recently as 1969, when the Arizona Game and Fish Commission designated it a protected species. However, Americans continued to covet jaguars as trophy animals and for their pelts, and they traveled to other nations to hunt them.

Since 1972, jaguars have had limited federal protection in both the United States and Mexico. In the United States in 1972, jaguars received federal protection under the Endangered Species Conservation Act, the precursor to the Endangered Species Act (ESA). When the ESA replaced this earlier act in 1973,

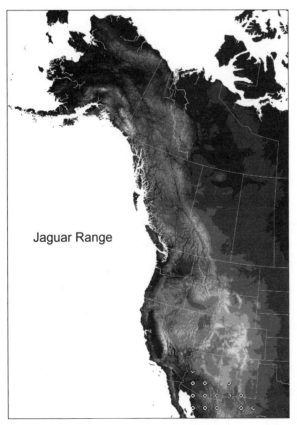

Jaguar Range

Figure 9.1. Map of jaguar distribution in 2013 within the Carnivore Way. (GIS map by Curtis Edson.)

a procedural oversight removed jaguar protection in the United States. At that point, the ESA only covered Mexican jaguars, since no US population existed. In response to concerns about the need to protect a potential US population of the species, in 1979 USFWS acknowledged its oversight, but still failed to expand jaguar protection.[16] In Mexico in 1980, jaguars received "endangered" status, with no regular hunting allowed legally. However, the federal government continued to issue special permits to hunt depredating jaguars as a way to cull them.

International jaguar protection dates to 1975, with the enactment of the Convention on the International Trade in Endangered Species (CITES). This law halted guided international jaguar hunts and trade in jaguar pelts. Mexico didn't sign this act until the late 1980s, so the jaguar pelt trade continued in that nation. But even with CITES protection in place, poaching and illegal trade continues to be a problem. In 2002 the International Union for the Conservation of Nature (IUCN) gave the jaguar "near threatened" status on its Red List of Threatened Species. Currently, jaguars have full protection across most of their range, with limited hunting allowed in Brazil, Costa Rica, Guatemala, Mexico, and Peru.[17]

In the United States, a series of jaguar sightings (e.g., Glenn's and Child's observations) prompted reconsideration of federal protection for this species. In 1997, in response to lawsuits, the US Fish and Wildlife Service (USFWS) listed the jaguar as endangered in Arizona and New Mexico. Also in 1997, USFWS convened the interagency Jaguar Conservation Team, known as the JagCT, led by the Arizona Game and Fish Department (AZGFD) and the New Mexico Department of Game and Fish. This interagency group had the primary task of figuring out how to protect jaguars and foster coexistence with them.[18]

It's unlikely that a breeding jaguar population exists in the southwestern United States. Since the 1990s there have been more than a dozen reliable (i.e., based on physical evidence) jaguar sightings in Arizona and New Mexico. In 1963 a hunter killed a female jaguar in Arizona's White Mountains, the last female of this species to be recorded in the United States. Most jaguar records since then have consisted of single males within 25 miles of the US-

Mexico border. Careful analysis of these and historical records suggests that a breeding jaguar population hasn't existed in the southwestern United States since the 1960s.[19]

South of the US-Mexico border, things look more promising. The Nacori Chico jaguar population, the northernmost breeding population of this species, lies in the Mexican state of Sonora, 130 miles from the border. This population contains an estimated 50 to 271 individuals. Since the 1990s, Sonora ranchers have killed many jaguars due to livestock depredation. Of them, several were females, including one with two kittens and another that was lactating. These deaths furnished evidence of a breeding population. However, this small population may be incapable of providing sufficient dispersing jaguars to recolonize suitable habitat in the United States in the near future. Further, we've seen that females of this species don't disperse long distances.[20]

In the Americas today, primary jaguar threats include habitat loss via deforestation and fragmentation, competition with humans for prey, and agricultural development (primarily ranching), which creates high potential for conflict due to cattle depredation. For example, Sonoran ranchers have removed thornscrub habitat to grow crops such as alfalfa for their cattle, thereby reducing jaguar habitat. To address these issues, eminent jaguar ecologist Alan Rabinowitz and others have initiated a strategy to create a north–south protected corridor for jaguar migration through the Americas. While working under the auspices of the Wildlife Conservation Society (WCS), Rabinowitz and Kathy Zeller developed a rangewide model of connectivity for this species. He's also written about jaguar conservation issues in his recent book, *An Indomitable Beast: The Remarkable Journey of the Jaguar.* Since 2008, Rabinowitz has been the CEO of Panthera, a science-based conservation organization dedicated to ensuring a future for the wild cats of the world, where he continues his corridor ecology work with Zeller.[21]

The region south of Arizona and New Mexico is critical for jaguar recovery in the southwestern United States. To get to the US, jaguars must travel through Sonora and Chihuahua, where they run a gantlet of threats to their survival and movement, from poaching to the habitat fragmentation caused by the US-

Mexico border fence. As we saw in chapter 3, this fence, built to reduce drug trade and illegal immigration, may be proving less effective at blocking human activity and more effective at blocking wildlife movements. But while the fence may be limiting jaguar movement across the border, the extent of this effect remains unknown due to the lack of radio-collar data for this species.[22]

The Ecological Effects of Jaguars

Like most large carnivores, jaguars can have strong effects on ecosystems. During an early 1930s bow-hunting trip to Sonora, Aldo Leopold wrote, "We saw neither hide nor hair of him, but his personality pervaded the wilderness; no living beast forgot his potential presence, for the price of unwariness was death. No deer rounded a bush, or stopped to nibble pods under a mesquite tree, without a premonitory sniff for *el tigre*. No campfire died without talk of him."[23]

Since Leopold penned these words, we've learned much about the ecological role of large carnivores. So just what is the jaguar's role? To begin with, it's an indicator species that can act as a surrogate for the integrity of the ecosystems it inhabits. Brian Miller and other conservation biologists also consider the jaguar an *umbrella species*—one that covers large areas in its daily or seasonal movements. Protecting sufficient habitat to ensure a viable population of such species benefits many others that are more restricted in their range. In the jaguar's case, protecting it can also create a protective "umbrella" to meet the needs of smaller species such as ocelots (*Leopardus pardalis*) and jaguarundi (*Puma yagouaround*).[24]

In terms of food web effects, this stealth hunter is unlikely to drive top-down trophic cascades (e.g., by influencing the behavior and density of its prey to the point that vegetation is released from herbivory pressure) due to its solitary nature and low numbers, particularly in the north. Its most significant ecosystem interactions occur via the food web subsidies it provides for other species, such as ocelots, coyotes (*Canis latrans*), bears (*Ursus* spp.), and birds of prey. Borderland jaguars typically kill one deer every three to five days, eating their fill and caching the rest. While jaguars sleep, other species opportunistically feed on the cached carcasses of jaguar-killed prey.[25]

Jaguars and People: The Macho B Case

Humans have been drawn to jaguars throughout the ages. Archeological evidence indicates that this animal played a prominent role in the religion, mythology, and art of Middle-American civilizations. The Olmec (1500 BCE), Zapotec, Mixtec, Mayan (400 BCE), and Toltec (650 CE) civilizations regarded the jaguar as the god of power and war. For example, the Mayans referred to the seat of power as the "jaguar mat," and going to war as "spreading the jaguar skin."[26]

Today, jaguars continue to represent wildness and power. Their allure is such that we've named a line of high-performance sports cars after them. Their conservation importance has inspired scientists to study them, gladly going nearly to the end of the earth (Argentina) and living in jungles filled with poisonous creatures (Belize). Indeed, this species has at times motivated people to go altogether too far in studying and conserving it.

Back in 1996, the Childs and Colvin jaguar encounter inspired them to found the Borderlands Jaguar Detection Project (BJDP). They used heat- and motion-sensitive cameras (called *camera traps*) programed to go off when an animal walked by in order to document the southwestern US jaguar population. Decades of hunting experience had given them intimate knowledge of the Arizona landscapes where people had spotted jaguars. Their perseverance paid off in 2001, when they captured their first jaguar image: a male whom they dubbed Macho A.

In 2004, graduate student Emil McCain began to work for the BJDP. A master's student in wildlife at Humboldt State University, California, studying cougars, McCain commuted between California and Arizona to pursue his passion for jaguars. With McCain's help, Childs was able to triple the number of camera traps deployed in various southern Arizona mountain ranges. Between 2004 and 2005, they baited some cameras with jaguar scat obtained from the Tucson zoo. McCain started participating in JagCT meetings. And beginning in 2006, he worked with BJDP field technician Janay Brun to set up and monitor camera traps.

The BJDP camera traps captured 15,000 photos, including 70 images of jaguars. Macho B, the same male Childs had photographed back in 1996

during his jaguar encounter, appeared in many of these 70 images, easily identified by the tell-tale Pinocchio-shaped rosette on his flank. Macho A disappeared
in 2004, with Macho B showing up on camera three hours later, precisely where
Macho A was last photographed. This led Childs to conclude that Macho B had
become the dominant male in the area. The cameras also photographed a third
jaguar, who bore a different rosette pattern than the others. But because after
Macho A's disappearance this other individual was never seen again, Childs and
McCain concluded that those images may have been simply the other side of
Macho A.[27]

These intriguing jaguar images raised interest in capturing and radio-collaring a jaguar in order to learn more about the conservation issues faced by
this species at the US-Mexico border. In 2008, AZGFD contracted McCain to
snare black bears (*U. americanus*) and cougars. He planned to use these species
as surrogates to model potential borderland jaguar movement corridors. Also
in 2008, $50 million in federal funds became available to study and mitigate the
negative effects on wildlife of the US-Mexico border fence.[28]

Since Macho A's disappearance in 2004, Childs and McCain hadn't picked
up additional individuals on camera. In 1996–97, the jaguar that Warner Glenn
originally sighted was rumored killed across the border in Mexico. In 2006,
Glenn's hounds treed a second individual in the same area as the first, but with
no subsequent sightings or camera-trap images. Thus, by 2008 USFWS concluded that Macho B was likely the only known jaguar in the United States. Environmental groups such as the Center for Biological Diversity (CBD) opposed
capturing Macho B, due to the stress this could potentially place on him given
his advanced age, and because losing him would mean losing the only jaguar we
knew of in the United States.[29]

In November 2008, McCain set a nonlethal trapline (a series of snares) for
his AZGFD black bear and cougar project, primarily in Arizona's Peñasco Canyon. McCain's field crew included AZGFD employees Thornton Smith and Michelle Crabb, plus BJDP employee Janay Brun. On February 4, 2009, McCain
picked up a camera image of Macho B taken two weeks earlier, one mile south
of Peñasco Canyon. The next day he discovered jaguar tracks that appeared to
have been made during the same period as the photo.

McCain and Smith reactivated the snares by opening them. To lure the male jaguar, McCain instructed Brun to bait one of them with jaguar scat from the Phoenix Zoo. Scat baits hadn't worked before to draw in Macho B, but Brun did as instructed. Jaguars are notorious for sustaining injuries when snared, as they fight very hard—harder than most species of large carnivores. McCain made the snares more jaguar-friendly by shortening the tether and taping the tightening point to prevent wire cuts to an animal's leg. He didn't have a permit to capture a jaguar, as this was prohibited under the ESA. Further, he was unqualified to handle such a delicate operation. His previous jaguar snaring experience had been in Sonora in 2004, where he caught two jaguars for radio-collaring, inadvertently causing the death of one due to improper handling.[30]

Once all was in place, McCain went on vacation to Spain, leaving the snares under the care of Smith and Crabb. Brun wasn't further involved in the capture. Smith and Crabb had never handled a jaguar before. Details of what followed are part of federal investigation records.[31]

On a frosty winter day two weeks after the snares were opened, Macho B stepped into the snare baited with jaguar scat. Smith reported Macho B's capture to AZGFD. By then sixteen years old, Macho B was the oldest jaguar ever documented in the wild. Although he was in very bad shape, he fought hard in the snare, sustaining a few minor cuts and abrasions and breaking a canine tooth at the root in the process.

Smith immobilized the jaguar with the drug Telazol. He and Crabb collected the usual biological specimens and data (blood, feces, weight, body condition). Neither of them had experience drawing blood from cats, so they only got a partial sample. They then mishandled the blood specimen by freezing it, thereby rendering it unusable. Macho B was hypothermic, due to the cold and the drug, so they covered him. When done radio-collaring the jaguar, they released him and waited for the drug to wear off. He stayed under for an extremely long time—six hours from start to finish. Most collaring efforts call for immobilizing an animal with drugs for just an hour or so. Video footage of the capture shows the jaguar wobbling away slowly, his hindquarters barely functioning.

In the days that followed, AZGFD issued congratulatory e-mails about Smith's efforts and a celebratory media release. Meanwhile, McCain, still in Europe, was receiving Macho B's GPS collar data on his computer. Within six days, Macho B's lack of movements indicated trouble. When a recently radio-collared animal stops moving for an extended period of time, something may be wrong (e.g., the animal is injured, sick, or dying). This can happen due to side effects from immobilization drugs, trauma from snaring, or many other factors. Eventually, twelve days after Macho B was captured, he was recaptured and transported to the Phoenix Zoo. A veterinary team from the zoo euthanized Macho B due to kidney failure. At the time of death, he hadn't eaten in over a week and was severely dehydrated.

Phoenix Zoo veterinarians and the University of Arizona (UA) Veterinary Diagnostic Laboratory became involved in Macho B's postmortem analysis. Zoo veterinarians took a culture from a tissue sample and diagnosed him as having severe subcutaneous emphysema, which develops when there is blunt trauma to the skin and air is trapped in a wound. He had a normal white blood cell count. The UA Veterinary Diagnostic Laboratory concluded that he was dehydrated and that his euthanasia may have been premature. Uncertainty exists about the cause of Macho B's death, because AZGFD had ordered only a partial, "cosmetic," necropsy, and because Smith and Crabb hadn't drawn usable blood at the time of his initial capture.[32]

Every tragedy has a hero, but this one has two. The first hero is Macho B, who demonstrated courage and strength in recolonizing territory at the northern edge of his species' range and then living long and well in a landscape dominated by humans. The second hero is Janay Brun.

Like Childs, Brun had an encounter with a cat that changed the trajectory of her life. In autumn 1999, she was hiking in Arizona with her dogs at dusk. At a bend in a trail at the mouth of a small canyon, she came upon a jaguar sitting on a granite boulder, silhouetted by the setting sun. As the jaguar leaped off the rock and onto the trail, it emitted a loud cough. It took off in the opposite direction from Brun, leaving her standing there, stunned, heart beating hard. At the time she had no way of knowing that Macho B—the animal on that trail—

would be the same individual she would one day study. She ended up working for the Borderlands Jaguar Detection Project for seven years, first as a volunteer, and then as an employee.

In discussing these events with her four years later, I found her still trying to understand the big picture of what we can learn from Macho B. In reflecting on his death, she says, "Macho B was lost in all of this—there was absolutely no concern about him. This was more about what people could get out of him."[33]

When Macho B stepped into that snare, he stepped into a web of environmental politics, border security, and quests for scientific funds. At stake was a $771,000 Department of Homeland Security (DHS) grant to study the effects of the border fence on jaguar movements. AZGFD reported his capture as "incidental," meaning that he'd been caught inadvertently in a snare set for a black bear or cougar. Guilt-ridden after Macho B's death, Brun divulged the truth to environmental reporter Tony Davis of the *Arizona Star*: that the snares had been set intentionally for Macho B and that his capture had been no accident.

In the investigation that followed, McCain was charged by the federal government with violation of the ESA. AZGFD disowned him, stating that he'd acted as an individual, without their knowledge of his actions. The Garrity Warning, a federal law that requires employees to answer questions with no counsel present in exchange for immunity from prosecution, protected AZGFD and USFWS employees involved in the Macho B controversy. The federal government prosecuted McCain and Brun for the capture and death of an endangered animal. McCain cast himself as a victim of the system. He stated that AZGFD had encouraged him to trap Macho B, in part to secure some of the DHS funds. He claimed that the agency had knowledge of his illegal actions and had helped him to conceal them.[34]

The victim in all this, besides Macho B, was Brun, who blew the whistle for moral reasons. When setting the snares, she asked McCain, her supervisor, about Macho B's age, health, and the safety of snaring him. Further, she asked McCain whether he had permits for everything. When he responded yes, she later told investigators that she proceeded to do as she was told as an employee.

And when she divulged McCain's wrongdoing, she didn't consider the personal repercussions of being a whistleblower.[35]

Lawsuits flew back and forth, but the bigger picture behind the web of intrigue that led to Macho B's capture was never addressed. Under the terms of a federal plea bargain, USFWS banned McCain from conducting research in the United States for five years. He's since begun Iberian lynx (*Lynx pardinus*) research in Spain. Brun was offered diversion, a legal process whereby charges are dropped upon reaching an agreement. For this agreement, she admitted that she was directed to place jaguar scat near a snare site in an effort to capture Macho B, that she was aware of his presence in the area, that the snares were set to capture cougars and bears, and that there was no permit to capture a jaguar (of which she was unaware until long after the act). After a few hard years, she's now living in Pennsylvania with her family, rebuilding her life and writing a book about her experiences.[36]

Jaguars and Corridor Ecology

Since giving the jaguar endangered status in Arizona and New Mexico in 1997, USFWS had failed to designate critical habitat for this species. Definition of *critical habitat* (i.e., the habitat essential to endangered or threatened species' survival) is necessary before a recovery plan can be drafted for a species protected by the ESA. Without a recovery plan, the jaguar was in conservation limbo: officially protected, but with no plan to move forward to restore it via a recovery plan. To address this lack, the Center for Biological Diversity sued USFWS in 2003. The agency responded by asking the JagCT (the jaguar conservation committee it had appointed) to conduct an analysis of *suitable habitat*. This is a legal and ecological term that refers to the habitat *potentially* suitable for use by a threatened or endangered species, typically expressed as "[species name] suitable habitat." The process of identifying specific existing suitable habitat for a species typically occurs as a precursor to defining *critical habitat* for that species, as part of a congressionally mandated process. In the southwestern United States, these analyses and the resulting models would create the foundation for federal jaguar recovery planning.

JagCT contracted James Hatten to do the Arizona analysis. Hatten and his colleagues used a Geographic Information System (GIS) to model jaguar suitable habitat by overlaying historic jaguar sightings on landscape and habitat features believed to be important, such as vegetation type, elevation, terrain ruggedness, proximity to water, and human density. They found suitable jaguar habitat in 21–30 percent of Arizona. Accordingly, they recommended that conservation efforts focus on rough, low-elevation scrub grasslands located near water, and on connecting that habitat to travel corridors within and outside Arizona in order to enable jaguar dispersal.[37]

The JagCT also contracted Kurt Menke and Charles Hayes to do a jaguar suitable habitat analysis for New Mexico. Menke and Hayes used reliable jaguar sightings in New Mexico to build a GIS model that included the landscape features known to influence jaguar habitat suitability. They characterized jaguar suitable habitat as having low human density, Madrean-type vegetation, water in close proximity, abundant prey (peccaries and deer), and rugged terrain (for concealment). The two areas with the highest probability of supporting jaguars were the Peloncillo and Animas Mountains in far southwestern New Mexico, and the Gila River and San Francisco River drainages farther north along the New Mexico-Arizona border. While their model could be used to evaluate potential corridors for jaguar travel, they recommended further analysis to identify potential linkages to habitats that currently sustain breeding populations of jaguars (e.g., the Mexican state of Sonora).[38]

In 2006, the Center for Biological Diversity (a JagCT member) created another habitat suitability model for New Mexico. Michael Robinson, who led this analysis, found a far broader area—nearly 50 percent of the state—suitable for jaguar recolonization. His model indicated that suitable habitat existed in most of the western half of the state, from the Mexican border to the state's northern border with Colorado. However, he used far more liberal parameters than the other modelers. For example, he expanded the designation of suitable habitat to include plant communities other than Madrean and thornscrub. Robinson noted that the biggest threats to jaguar conservation were busy human-travel corridors, such as Interstate highways.[39] Neither Robinson's nor any of the other jaguar habitat models referenced the border fence, because it didn't exist yet.

Later in 2006, in further response to the 2003 CBD lawsuit regarding critical habitat, USFWS made a statement that the jaguar didn't need special protection in the United States to survive. This prompted another lawsuit on the basis that the ESA wasn't simply about species survival, it was about species *recovery*. Due to the Macho B incident and the questions it raised about whether jaguars even exist in the United States, and if so, how we could better protect them, a federal judge ordered USFWS to create a proposal for jaguar critical habitat.

In 2011, the federal government began to investigate the ecological impacts of the border fence on jaguars. The Department of Homeland Security awarded researchers from the University of Arizona's Wild Cat Research and Conservation Center a $771,000 grant to study this corridor. This project's main objective is to help define jaguar critical habitat in a way that incorporates the effect of the border fence. Lisa Haynes and Melanie Culver head the research team, which includes Jack Childs. To achieve their objective, they're monitoring jaguar movements along the border with 240 motion-activated cameras set on wilderness trails. In 2012, the cameras recorded several images of a male jaguar in the Santa Rita Mountains, 40 miles southeast of Tucson. These images provide evidence that other jaguars besides Macho B are using this area.[40]

In early 2012, USFWS assembled a binational Jaguar Recovery Team, co-led by Carlos Lopez Gonzalez and Howard Quigley and composed of Mexican and US experts. The team focuses on the Northwestern Jaguar Recovery Unit, which encompasses much of the Sierra Madre Occidental in western Mexico, including management units in Sonora, southeastern Arizona, and extreme southwestern New Mexico. They're identifying jaguar recovery needs throughout its range and the role of the northwestern population in the conservation of the whole species. Additionally, they're examining the importance of habitat connectivity.[41]

In April 2012, USFWS reversed its 2006 decision and filed a proposal to designate 800,000 acres in Arizona and New Mexico as jaguar critical habitat. One-eighth of this lies in the Peloncillo Mountains, on the Arizona–New Mexico border. Critical habitat also includes the Baboquivari Mountains in southeastern Arizona and New Mexico's southwestern "bootheel," and another group of mountains known collectively as the Sky Islands. An archipelago of 52 mountains separated by valleys and grasslands, the Sky Islands lie at the nexus of several desert and forest biomes (e.g., Chihuahuan Desert, Sonoran Desert, Southern Rocky Mountains, and Sierra Madre Occidental). Ecologists consider the Sky Islands a biodiversity "hotspot" and a key area for wild felid conservation. In defining critical habitat, USFWS evaluated previous habitat models and created a new model, in consultation with the Jaguar Recovery Team. Criteria were similar to those identified by Menke and Hayes, as well as Hatten and colleagues.

In its proposal, USFWS argues that the northern jaguar population, while peripheral to those in Mexico and farther south, is essential to the species' survival because this northern population has adapted to different environmental conditions than southern populations. This evolutionary difference can make the species more resilient to climate change. Further, according to Quigley, the agency owes this species critical habitat, because the animal's extirpation occurred largely at the hands of the government via the federal predator-control program in the United States. To that end, they recommend a reserve network that would maintain ecological connectivity via jaguar population cores and corresponding dispersal corridors.[42]

USFWS's jaguar critical habitat proposal has reopened a lively debate on this topic. On one side of the argument, Rabinowitz considers critical habitat designation in the southwestern United States unwarranted because the area in question has always been marginal habitat at best. The absence of US Native American jaguar iconography furnishes evidence of this. While he agrees that peripheral populations are crucial to a species' conservation, this only applies to breeding populations. The southwestern US jaguar population, if it can be deemed a formal population rather than a loose group of transient animals, doesn't constitute a breeding population. Rabinowitz interprets the lack of a jaguar breeding population in the southwestern United States as further evidence that this habitat is no longer ideal for the species. Given the limited federal funds for jaguar recovery, he questions whether designating critical habitat in the north would be the best use of these funds.[43]

In contrast, the Center for Biological Diversity is requesting a broader designation of critical jaguar habitat. This would include expansion of critical habitat into the Mogollon Rim, the Chiricahua Mountains, and Coronado National Forest north of US Interstate 10. CBD asserts that a southwestern US jaguar population can provide resilience for other populations (e.g., Sonoran jaguars). Full recovery of this species north of the US-Mexico border could entail maintaining critical habitat linkages (corridors for dispersal) and possibly implementing a reintroduction and augmentation program similar to the Mexican Gray Wolf Recovery Program.[44]

In reflecting on jaguar critical habitat ten years after he and Hayes created their model, and given what we now know about climate change that we didn't know back then, Menke has some interesting insights. According to Menke, cold is a limiting factor for this species. However, as the climate continues to warm, jaguars will be expanding their range northward. By 2100, if climate change continues at its current pace, wild or undeveloped lands in central and northern New Mexico and Arizona may become important jaguar habitat and could provide refuges for this species. In the meantime, Menke points out that the lack of knowledge about jaguars in Sonora and the southwestern United States reinforces the need to maintain connectivity for this species in the southwestern United States.[45] Maintaining connectivity would include addressing any problems created by the border fence.

The key to jaguar conservation lies in working at multiple scales. A metapopulation (i.e., population of populations) approach means conserving local jaguar populations in Arizona and Sonora. While DNA analysis shows that genetic connectivity occurs between jaguar populations throughout the Americas, it's important to keep dispersal corridors open. This means working across boundaries such as the US-Mexico border. At the metapopulation level, resilience can be evaluated via effective movements of juvenile animals away from their natal home range to find a mate and establish a new territory. Based on its natural history traits (e.g., ability to make long movements, low reproductive rate, need for cover, broad diet), the jaguar is moderately resilient. Failing to conserve fringe populations of moderately resilient species could lower their ability to adapt to environmental changes such as global warming.

Reserves are part of the solution. In the 1980s, Alan Rabinowitz established the world's first jaguar reserve, in Belize's Cockscomb Basin. Since then

Cockscomb jaguar research has blossomed, supported by multinational funding from as far north as Canada.[46] The Northern Jaguar Reserve, established in 2008 in Sonora by a consortium of binational nongovernmental organizations, such as Naturalia, is manifesting this sort of landscape-scale conservation vision.

The Northern Jaguar Reserve comprises 50,000 acres in the Sonoran Sierra Madre foothills, in prime jaguar habitat. Stitched together from two former cattle ranches, Rancho Los Pavos and Rancho Zetasora, the Northern Jaguar Reserve contains all the elements jaguars need in order to thrive: deep canyons and valleys, rough mountain terrain, abundant water, plentiful prey, and few people. Carlos Lopez Gonzalez and other scientists have been conducting research here on the jaguar's prey base, population dynamics, dispersal, and human coexistence with this powerful felid. Additionally, between the reserve and the US-Mexico border lie 250,000 acres of permanently protected land, including private ranches and the Ajos-Bavispe National Forest Reserve and Wildlife Refuge. These ecological stepping stones can provide the "missing link" in jaguar conservation at the northern extreme of this species' distribution.[47]

Coexisting with Jaguars

As is the case with other large carnivores in North America, the best hope for jaguar recovery lies in maintaining connected habitat and mitigating conflicts with humans. A conservation strategy that shifts US federal funds south of the US-Mexico border would help to do that. To that end, in 1999, the Wildlife Conservation Society (WCS) convened 36 jaguar biologists from eighteen nations to identify all known areas inhabited by jaguars. Their resulting analysis, which included mapping the spatial extent of jaguar knowledge and the status of existing populations, enabled them to determine that, for borderland jaguars, what happens in Sonora is critical to their long-term survival. But beyond such insights, the importance of this WCS effort lies in the understanding that saving jaguars requires international, rangewide planning that prioritizes efforts based on ecological importance.[48]

On a grassroots level in Sonora, thanks to scientists Octavio Rosas-Rosas and Raul Valdez, some very proactive coalitions are emerging. Since the 1600s, the ranchers here have seen themselves as living in conflict with large predators. Jaguars prey on cattle, but to what extent that had been occurring had never been measured. We've seen that the Sonoran Nacori Chico jaguar population provides a key source of jaguars dispersing into the southwestern United States. However, since the Mexican federal government implemented jaguar protection in 1980, Nacori Chico ranchers have taken predator control into their own hands—illegally. A high level of conflict between jaguars and ranchers inspired Rosas-Rosas and Valdez to focus their research here.

Rosas-Rosas and Valdez began by helping eight Nacori Chico landowners form an alliance. This group agreed to voluntarily suspend illegal predator-control programs if they received compensation for cattle losses. The researchers then helped ranchers work with a private outfitter to initiate conservation hunts of white-tailed deer on their lands. A portion of this income (approximately $30,000 per year) went to the ranchers to compensate them for cattle losses. By involving landowners in a democratic and voluntary conservation program, Rosas and Valdez were able to significantly reduce jaguar killing. Additionally, they confirmed that jaguar and cougar depredation accounted for less than 14 percent of total cattle losses. Their research provides an example of how work that responds to ranchers' concerns and needs can help jaguars survive.[49]

Conservation-minded ranchers have tried jaguar-friendly ranching independently of coalitions. One such rancher, Carlos Robles Elias, owner of Sonora's Rancho Aribabi, located along the Cocospera River, 30 miles south of the US-Mexico border, has attempted to turn his family ranch into a working refuge for biodiversity—a place where jaguar conservation takes precedence over running cattle. His grandfather purchased El Aribabi in 1888. Robles grew up in Nogales, Sonora, where his father ran restaurants. An avid naturalist as a boy, he moved to Rancho El Aribabi at the age of 27 with his wife and young son; and in 1987, he took ownership of one-third of the ranch—10,000 acres.

Robles' conservation efforts on El Aribabi developed over several decades. In the 1990s, he began to work with biologists to monitor bird communities and restore riparian areas on his land. Multinational conservation organizations such as Joint Venture helped in this endeavor. He participated in Mexico's UMA program, which encourages ranchers to create practical plans for sustainable wildlife management (including hunting). In 2000, he started re-ducing the number of cattle on the ranch (eventually reaching zero). In 2006, he partnered with the Sky Island Alliance (SIA), a grassroots conservation or-ganization that aims to protect and restore the native species of this region. He adopted a no-kill predator policy, and SIA set up camera traps on his land. The camera traps began to pick up images of jaguars and the northernmost breeding ocelot population. In partnership with SIA, Robles began to run conservation deer hunts and ecotourism activities. Tracker and wildlife biologist Cynthia Lee Wolf, who's worked on jaguar and wolf conservation with ranchers in the United States and Mexico for decades (including the Borderlands Jaguar Detec-tion Project), helped guide Robles' ecotourism program.[50]

Robles, who needed the ranch to sustain itself financially in order to sup-port his family, intended ecotourism to replace income lost from reducing his cattle operation. However, while drug-cartel violence along the US-Mexico border didn't affect El Aribabi directly, fear of such violence effectively reduced ecotourism there by over 60 percent. In 2011, lacking direct funding from SIA or other conservation organizations, Robles was forced to resume cattle ranch-ing at high stocking levels (approximately 1,000 head). That year, the Mexican government designated Robles' ranch a "protected natural area," the highest conservation award given to private land. While this raised his conservation profile, it didn't lead to additional funding. In further recognition of his con-servation efforts and expertise, USFWS appointed him to the Jaguar Recovery Team in 2012.

Robles' experience illustrates that, in today's economy, conservation ranch-ing isn't always self-sustaining. According to Wolf, government incentives, sup-port from both nongovernmental conservation organizations and from donors support is essential. Robles' situation is particularly unfortunate, because his ranch represents an ideal stepping-stone for jaguar dispersal—and a premier safe haven from predator control.

Macho B taught us painful conservation lessons that can help advance carnivore conservation. He taught that actions taken without proper peer-review and prior approval by endangered-species recovery committees can lead to unnecessary death of animals, as happened with Macho B. Additionally, a

Earth Household

In this book we've explored the Carnivore Way by following the footsteps of the large carnivores. We've seen how the animals whose stories I've shared are teaching us the importance of coexisting with them and the wildness they bring to this earth. Their lessons are really important because, right now, our world is at greater risk than ever. In the Anthropocene epoch, our growing human population, need for food and fuel (e.g., natural gas), and the warming planet are creating a hemorrhage of extinction. We're losing species at the rate of 6 percent per decade, primarily due to human degradation of ecosystems.[1] Large carnivores, which touch everything in the web of life, create biodiverse, healthy ecosystems. Richer in species, ecosystems that contain large carnivores will be more resilient to change, and therefore will better enable humans to live sustainably, capably, and happily on this planet.

To begin with, large carnivores are teaching us that they need lots of space, and that national parks aren't enough to meet their fundamental needs for food and a mate. During an eighteen-month period, Pluie, a young radio-collared wolf (*Canis lupus*), traveled an area that encompassed more than 40,000 square miles, crossing more than 30 legal jurisdictions, including two Canadian provinces and several US states. In her hegira, she showed us that carnivores need connected landscapes that transcend political boundaries, and that thinking about conservation on a continental scale is essential for them and other wild animals in order for us to help them maintain their genetic diversity and resilience. Pluie inspired conservationists and scientists to find continental-scale solutions, using the science of corridor ecology.[2] Other animals, such as bear 64 in Banff National Park, have taught us that solutions such as the Bow Valley wildlife crossings really do work. The Highway 1 overpasses there enable bear 64, her cubs, and other animals to find safe passage over four lanes of heavy vehicle traffic, thereby linking wildlife populations that this road had fragmented.

Large carnivores are showing us that their wildness serves a real ecological purpose. Throughout the Carnivore Way, wolves, lynx (*Lynx canadensis*), and other carnivores are teaching us about how rewilding nature by bringing them back from the brink of extinction is as necessary to the exchange of energy in food webs as the bottom-up flow of soil nutrients. By creating trophic cascades—top-down food web relationships in which apex predators affect their prey, which in turn affects the prey species' consumption of food—wolves and other apex predators increase ecosystem health. For example, lynx prey on snowshoe hares (*Lepus americanus*), which eat willows (*Salix* spp.). By reducing snowshoe hare numbers, lynx indirectly create healthier willows, and these healthier willows can provide better habitat for songbirds. But in addition to the actual, direct acts of predation, these predator–prey relationships can have indirect effects related to fear. In Glacier National Park, for example, fear of wolves makes elk (*Cervus elaphus*) more alert. To avoid getting attacked by wolves, elk must move around more and avoid spending time in aspen (*Populus tremuloides*) stands that have burned and become filled with deadfall. The downed trees in these burned stands make it difficult for elk to run to escape wolves. In Glacier, burned stands now contain thickets of aspen saplings growing vigorously above the reach of elk. Thus, together with fire, wolves are helping aspen flourish.

The animals in this book are also teaching us that without connected corridors and adequate regulatory mechanisms, they may go extinct. That we have any carnivores left at all today is due to the web of environmental laws that protect them. This imperfect web provides the framework for wildlife conservation across the three nations in the Carnivore Way: Canada, the United States, and Mexico. In Canada, the Species at Risk Act (SARA) protects species from extinction. However, Canadian citizens have very limited ability to hold their federal government legally accountable for this law's enforcement. In the United States, we have the Endangered Species Act (ESA), one of the most powerful environmental laws in the world due to its substantive nature. US citizens dissatisfied with this law's implementation can sue the government. However, the ESA is cumbersome and expensive to apply. For example, as of 2013, it had taken one and a half decades to create a lynx recovery plan, two decades to protect the wolverine (*Gulo gulo*), and four decades to fully protect the jaguar (*Panthera onca*) and create a recovery plan. While successes such as the wolf in the Northern Rocky Mountains (NRM) demonstrate that this law works, we have failures, such as the case of the Mexican wolf (*C. l. baileyi*) in the southwestern United States, which, due to illegal killing by humans, hasn't reached the modest recovery threshold of 100 animals despite almost two decades of conservation efforts. Of all three nations, Mexico has the weakest environmental laws. But even there, people are making headway in conserving carnivores, such as the jaguar, with government support of private land stewardship. Tribal law sustains the rich cultural heritage and traditions of Canadian First Nations, US Native Americans, and Mexican indigenous people. In the US and to a lesser extent in Canada, Native Americans and First Nations, which have legal sovereignty as nations per the US and Canadian federal governments, are also wrestling with modern natural-resources problems, such as grizzly bear (*Ursus arctos*) depredation on livestock calves and energy development on their lands. In Mexico, since the aboriginal people do not have standing there, they are not as actively involved in natural-resources management from a federal perspective.

A single, vibrant thread runs through all these lessons about connectivity, food web relationships, and environmental law: *coexistence.* Carnivore conservation is about coexistence, defined as our human ability to share landscapes with big, fierce animals that are sometimes our competitors for food. In or-about them from an ethical perspective. And therein lies the source of all of the debates about carnivore management described in this book, from grizzly bears to wolverines to jaguars.

Nelson and Vucetich cite two famous conservationists, Gifford Pinchot and Aldo Leopold, to illustrate the spectrum described above. Pinchot, named the first chief of the US Forest Service in 1905, created the concept of wise use of resources. He wrote, "There are just two things on the material earth, people and resources." Here he was referring to the *instrumental value* of nature, as something that exists purely for human use and exploitation. Between 1910 and 1948, Leopold, who created the science of wildlife biology, was among the first to recognize that carnivores matter due to the ecological benefits that accrue from them. His seminal land-ethic statement—"A thing is right if it tends to preserve the integrity, stability, and beauty of the biotic community; it is wrong otherwise"—addresses the middle of the spectrum (that living things have value because they create healthier ecosystems). Further, he stated that "A land ethic changes the role of *Homo sapiens* from conqueror of the land-community to plain member and citizen of it. It implies respect for fellow members, and also respect for the community as such." This can be interpreted as meaning that all living beings have intrinsic value, and so they should be conserved. Leopold's words spurred the environmental movement of the 1960s, which begat environmental laws such as the ESA.[4]

Several factors have shaped our relationship with wildlife in America. First, in the sixteenth century, Europeans—most of whom hadn't been allowed to hunt freely in the nations from which they originated—colonized what is now Canada, the United States, and Mexico. During European colonization, humans depended on hunting wild meat for sustenance far more than we do today. And in settling the New World, one of the first rules Europeans established was that, unlike in Europe, where often only royalty and landed gentry had the right to hunt, in North America, everyone should have free access to hunting. Furthermore, everyone should be allowed to bear firearms. These democratic ideas, eventually became the foundation for managing wildlife as a renewable, harvestable resource in all three nations. However, giving European settlers the unrestricted right to hunt resulted in a hunting free-for-all that lasted 400 years and caused widespread ecological damage.[5]

Lacking hunting regulations (e.g., how many animals could be killed, for what purpose, and when), and perceiving North America as a limitless land of plenty, we proceeded to hunt many species to extinction. *Market hunting*, defined as hunting wild animals in high volume for commercial gain, eliminated many species. As settlers advanced westward, they left a wake of extinction, including elk in the Midwest, bison (*Bison bison*) on the plains, beavers (*Castor canadensis*) in the Rocky Mountains, and large carnivores just about everywhere. In 1887, this wildlife plunder inspired Theodore Roosevelt, George Bird Grinnell (who founded the first Audubon Society), and others to create the Boone and Crockett Club, a hunter/sportsmen's organization. They wanted to halt market hunting and implement ethical principles for wildlife conservation that included hunting for recreation and sport. And in very short order, Boone and Crockett Club members, who were men of influence, created and passed some of the first US environmental laws. This included the Lacey Act of 1900, which prohibited the interstate transport of birds and other wildlife. This law ended market hunting, because people were no longer able to trade and distribute large amounts of dead game or their byproducts, such as feathers. By the 1910s, hunting as a conservation tool, which meant managing wildlife to increase the amount of harvestable game (by improving habitat and killing predators), had become the norm.[6]

In 2001, Canadian wildlife ecologist Valerius Geist and his colleagues created the North American Model of Wildlife Conservation. Based on principles in place since Roosevelt's era, the North American Model represented the first time these ideals were articulated in an integrated manner. Today, state and provincial wildlife management agencies strongly adhere to the North American Model.[7] It contains seven principles:

1) wildlife is a public trust (i.e, all wildlife are owned by citizens and managed for them by the government);
2) markets for game should be eliminated;
3) wildlife should be allocated by law (via hunting regulations);

4) wildlife should only be killed for legitimate purposes;

5) wildlife is an international resource;

6) wildlife should be managed with science-based policy; and

7) hunting should be a democratic process, with everyone having free access to wildlife.[8]

Working from these tenets, we can gauge successful conservation of a wildlife species by our ability to hunt it.

In every chapter of this book, we have looked at how management practices, be they creating connected corridors across roads for wildlife, setting hunting policy for cougars (*Puma concolor*), or using trophic cascades concepts to restore forests, are being applied in order to conserve carnivores. We have seen big successes, such as wolf recovery in the NRM, but we have also seen setbacks, such as the current policy of hunting NRM wolves (which no longer have ESA protection) down to the lowest level possible, using North American Model-based state wolf-management plans. This example demonstrates that for carnivores to truly recover, we need to rethink the North American Model.

Nelson and his colleagues suggest that the North American Model contains flawed logic. It attributes wildlife conservation entirely to hunters. Geist and others point out that we have wildlife today because hunters, such as Roosevelt, created a conservation vision to sustain wildlife. Others in the hunting community agree that hunting has led to the development of environmental virtue. However, the North American Model doesn't acknowledge the huge contributions to wildlife conservation that have come from outside the hunting community. For example, John Muir, who in the late 1800s advocated preserving nature, changed the way Americans envisioned their relationship to the natural world. This shift caused people to support protecting landscapes such as the Yosemite Valley and the animals that made their homes there.[9]

In the North American Model, Geist and his colleagues argue from a purely instrumental perspective that wildlife should be managed ethically, like a crop, so that human hunters reap the benefits. Nelson and others indicate that the North American Model doesn't specify how this is to be achieved, which means it can be used to justify wolf slaughter via aerial gunning in order to increase huntable caribou (*Rangifer tarandus*), for example. The North American Model calls for science-based management, but whose science is "best science"? It calls for the end of market hunting, but doesn't acknowledge that hunting today is a highly commercialized endeavor (e.g., $20,000 guided grizzly bear hunts). Finally, the North American Model focuses solely on conserving wildlife for hunting purposes. Yet in the United States, only 6 percent of citizens hunt. I'm part of that 6 percent. But in a democracy, basing wildlife management on a hunting paradigm that doesn't address the needs or desires of the non-hunting public (the vast majority of Americans) is wrong if, as the North American Model states, all wildlife is owned by all citizens jointly.[10]

In any discussion about large carnivore conservation with managers at federal and state levels, the North American Model comes up. The purpose of large carnivore conservation, managers say, is to recover a species like the grizzly bear so that we can hunt it. Yet recovering and conserving the grizzly bear, which has the lowest reproductive rate of any large mammal, and then allowing it to be hunted, is wrong ecologically. And in light of what we know about trophic cascades, recovering an animal like the wolf, removing its ESA protection, and then hunting it down to a bare legal minimum (e.g., ten breeding pairs per state in the NRM) is also wrong scientifically. And if you believe grizzly bears and wolves have intrinsic value, then you also could argue that such carnivore-management strategies are unethical.

Science tells us that if we want to have a healthy world, we need to conserve the carnivores.[11] Doing so necessitates creating a new model for coexistence. This means developing a contemporary land ethic, one that takes us beyond the North American Model and seeing wildlife as a crop or a renewable resource.

Leopold died unexpectedly of a heart attack at the age of 61. In the posthumously published *A Sand County Almanac*, he presents the land ethic—essentially a fledgling environmental philosophy. He'd probably planned to spend another twenty years refining his ideas about how humans could live more rightly on this earth. Instead he left us a powerful, albeit incomplete, roadmap for moving forward with conservation.[12]

We'll never know what Leopold would do today, given the threats we face that were unknown in his era, such as climate change. However, a contemporary land ethic would flesh out the bones of Leopold's ideas with what we now know about how the world works (corridors, trophic cascades) and how rapidly we're losing life on our planet, especially native species and taxa such as carnivores and pollinators that help create healthier biota. Such an ethic would provide a more inclusive way of coexisting with nature—one that leads to healthy eco-systems and thriving populations of animals, including carnivores, with which humans would live ethically and peacefully.[13] Such an ethic wouldn't preclude hunting, for humans have hunted animals for food since the early Paleolithic period (2.6 million years ago), and Leopold himself was an avid hunter. According to ethicist Charles List, a new wildlife conservation model would mean redefining hunting as a practice in which we don't treat animals like a crop to be harvested, and in which we exercise more restraint and respect for the animals that we hunt. In creating such a model, we'd consider not indiscriminately hunting carnivores.[14]

Is this vision of a more ethical relationship with carnivores attainable? To find out, I needed to journey deeper into wildness than I'd gone before, to a place where people are living very peacefully and respectfully with bears and wolves.

CB ED

It took four flights on progressively smaller aircraft, the last one a seaplane, plus two trips on small boats, the last one a sailboat, and six days to travel the 900 miles as the raven (*Corvus corax*) flies to the Great Bear Rainforest in coastal British Columbia from my northwest Montana home. For much of this journey there were no cell phone or Internet connections, and—more importantly—no roads.

I ventured into the Great Bear Rainforest to learn new lessons about coexis-tence. This meant going there in autumn, at the height of the salmon run, when pinks (*Oncorhynchus gorbuscha*), chums (*O. keta*), Coho (*O. kisutch*), and sock-eye (*O. nerka*), their bodies battered, bloody, and egg-heavy, instinctively find their way back to the coastal streams in which they were born. They lay their eggs in gravel beds and die, providing a feast for the bears (*Ursus* spp.), wolves, and bald eagles (*Haliaeetus leucocephalus*) that show up in droves to feed on this bounty, as they have throughout the ages. Word was that humans in this place had returned to a much older way of coexisting with carnivores. Indeed, Canadian bear expert Charlie Russell had come here to spend several years living with wild bears.

This remote Canadian temperate rainforest covers 21 million coastal acres from the north end of Vancouver Island to the British Columbia–Alaska border. Mostly mantled by nearly impenetrable western red cedar (*Thuja plicata*), western hemlock (*Tsuga heterophylla*), and Sitka spruce (*Picea sitchensis*) forests, 5 million acres of this ecoregion are protected. These discontinuous *conservancies*, as they're called, resulted from a collaboration between the eighteen First Nations who have territories in this ecoregion, the Nature Conservancy, Pacific Wild, conservationist Ian McAllister, and the provincial government. These sanctuaries are closed to logging, mining, gas-well drilling, and the hunting of bears and wolves. Since a prime tree here can be worth $10,000 when cut down, such protection is no small feat. Only accessible by boat or seaplane, these conservancies lie scattered like emeralds flung from a giant's fist across 250 miles of ragged coastline.

I sailed into this rainforest on the Ocean Light II, a 71-foot ketch, with Jenn Broom, the boat's owner, as well as Captain Chris Turloch and biologist Jim Halfpenny. My other companions included my husband Steve and several friends. Chris expertly navigated the boat deep within a maze of fjords. Breeching humpback whales (*Megaptera novaeangliae*) and killer whales (*Orcinus orca*) relentlessly hunting Steller sea lions (*Eumetopias jubatus*) filled these coastal waters. Along wave-scoured islands, amid the spill of sea foam across curving shores, and in estuaries where the mouths of rivers met the sea, we looked for bears—and found them.

Both grizzly bears and black bears (*U. americanus*) live in this rainforest. They partition themselves spatially by species, to avoid conflict. Grizzly bears only inhabit some islands, and black bears only inhabit others, which makes life simpler. This area also contains the largest population of Kermode bears (*U. a. kermodei*)—a very rare subspecies of black bear also known as the Spirit Bear, due to its white fur. Approximately 400 Kermode bears live here.

On the sixth day of our journey, Chris took us to Khutze Bay, a grizzly bear–dominated estuary. We set anchor and spent the night, the steady patter of rain on the deck lulling us to sleep in our cozy berths. At first light, we set out from the sailboat in a small Zodiac inflatable boat. Keeping the outboard engine purring on low throttle to avoid disturbing the wildlife, we cruised slowly up the Khutze Inlet, seeing bald eagles and gulls everywhere. Chris stopped the boat against a seagrass-covered shore, and waited. It didn't take long.

Two figures emerged from the mist, walking toward us: a huge, beautiful grizzly bear mother and her tiny cub of the year. They stopped at the shore, 30 yards from us, where the bear mother immediately got down to business. She put her face into the water and swiftly pulled out what must have been a twenty-pound Coho salmon, its silver body flopping around, and passed it to her cub. The cub chewed on it a bit, and then dropped the slippery fish. She found the cub two other equally hefty salmon, which the cub promptly fumbled. The third one threw the cub off balance, and both salmon and cub ended up in the water, where the fish swam off.

Before joining her cub in the water for more fishing lessons, the bear mother's eyes met mine, mother to mother, a classic "What's a mother to do?" expression on her face. I know how it is, I thought, returning her look—I'm a mother too. In her body language, especially in the relaxed eye contact she made, this was a *very* different bear from any I'd experienced in Katmai and my wild Montana home. And she prompted me to think about what coexistence really means.

Until now, my hundreds of meetings with bears, while peaceful, typically consisted of both of us consciously trying to minimize conflict. These accidental meetings often occurred while I measured aspens, or walked in the forest on my land. In these meetings, the bears usually didn't face me, instead passing me in profile without overtly acknowledging my presence, the better to avoid trouble. The Khutze bear mother brought up things beyond the pale. Was coexistence once very different from how we experience it today? Was the calm communion I experienced with her anything like the way our relations with bears once were? And if so, when did our relations with bears and the other carnivores go so wrong?

On the eighth day, we made our way to Gribbell Island, a 70-square-mile islet between the Inside Passage and the Douglas Channel. This lush, wedge-

Figure C.1. Grizzly bear mother in Khutze Inlet, Great Bear Rainforest, British Columbia. (Photo by Cristina Eisenberg.)

shaped island contains a lot of bears and wolves, but no human settlements. In fact, it provides home to the largest known concentration of Kermode bears.

The only way that one can set foot on Gribbell Island is with the permission of the Gitga'at, a Tsimshian First Nations band that controls access to this land. The Tsimshian live in coastal British Columbia, along mainland inlets and estuaries, and on islands. In their mythology, they and the other First Nations in coastal British Columbia consider themselves related to bear, wolf, raven, and killer whale, and don't hunt these animals. So it makes perfect sense that today the Gitga'at serve as gatekeepers to one of the most ecologically sensitive bear sites in North America.

Lead Gitga'at guide Marven Robinson met us as we clambered out of the Zodiac at low tide onto a slick-rock intertidal slope dappled with sea stars the color of pink grapefruit. Sturdily built and in his mid-forties, he wore chest-high waders and rain gear, a baseball cap on his head. He led us through an opening

Figure C.2. Bear mask carved out of western red cedar by Haida artist Jimmy Jones. (Photo by Cristina Eisenberg.)

in an alder (*Alnus rubra*) thicket, along a narrow, fern-edged trail toward the island's wild core. He'd grown up in this rainforest, in the nearby mainland village of Hartley Bay (pop. 200), which is accessible only by sea, so he knew this island and its bears intimately. As a child, like others of his band, Marven was brought up to respect bears and not hunt them. From the mid-1800s to the late 1990s, outsiders came to Gribbell Island and the surrounding region to hunt bears—the bigger the better. They especially came for trophy spirit bears hunts. Two decades ago, Marven was one of the people instrumental in getting this place and its bears protected. These days he guides people into Gribbell Island to raise awareness of bear conservation.

In a light rain, we followed the trail along the crest of a ravine that paralleled a narrow creek. Mist fingered the cedars, spruce, and alders that grew right up to the stream in this verdant rainforest. Moss clung to every surface. Raven gronks echoed through the forest nave. Faint wolf howls came from far upstream. About one mile up the trail, Marven took us down through the woods to the stream, to a viewing platform suspended on stilts, where we could sit out of the rain and watch bears. However, he had other plans for me. He took me down to the creek, below the platform.

"Sit here," he said, "and wait." Then he left.

I sat and waited. I don't know how long I waited. This primeval creek felt like a place out of time. Over the sound of creek water flowing on rocks, I heard the wanton cries of pileated woodpeckers (*Dryocopus pileatus*) marking their turf, and the husky chitter of pacific wrens (*Troglodytes pacificus*) foraging in the understory. The water churned with salmon bodies, mostly gleaming Cohos and dun-colored pinks, their humped backs protruding above the water's surface. Occasionally an electric-red sockeye finned upstream through the Medusa tangle of other species of salmon, all fighting the current to lay their eggs. Dead salmon, mission accomplished, littered the streambanks and lay washed up on rocks. From time to time, ravens swooped down from their perches to scavenge these dead fish.

Eventually a figure emerged from the woods, like an apparition. The big, black bear steadily approached the streambank, moving as if in slow motion. I sat there, spellbound. That bear, the local dominant male, was called Scarface. Battle-scarred (hence his name) and still in his prime, his body rippled with muscle and fat. Moving gracefully for such a large animal, he stopped about fifteen feet from me.

Chris talked to the bear. "Good to see you," he said. "You've been eating well."

The bear turned to face me and looked at me with soft, nutmeg-colored eyes. And in those long moments when our eyes met, there was no "us" and "them." Just "we"—two living beings here on this creek at the height of the salmon spawn, meeting in peace. Then he broke our gaze, looked intently into the fish-filled water, opened his massive jaws, and lunged faster than I thought a creature his size capable of moving. In less than two of my fast heartbeats, he pulled out a 30-pound Coho salmon. He sat there and calmly devoured the fish: first the roe, then the firm flesh, followed by the head. After he finished, he looked at me again, his eyes still soft, a crimson gobbet of salmon meat clinging

Figure C.3. Black bear fishing at creek on Gribbell Island, Great Bear Rainforest, British Columbia. (Photo by Cristina Eisenberg.)

to his chin. Then he turned away and departed the way he'd come, walking with loose-limbed grace atop a sway-backed mossy red cedar log that had fallen into the creek long ago. He jumped off the log and vanished into the forest understory, the alders and salmonberry bushes (*Rubus spectabilis*) rustling with his passing.

I sat there for a long while, unable to speak, taking in deep lungfuls of rainforest air and the comingled scents of wet cedar and rotting salmon—life and death. Marven brought people here, to this ursine inner sanctum, because he believed that we need to look into these bears' eyes to feel that connection, living being to living being, in order to want to do something to protect them from threats such as a proposed gas pipeline.

The bear mother and Scarface had shown me how it once was, between us and wild creatures sharp of tooth and claw, long before we thought we knew everything and could grow forests and elk like we grow cabbages (to paraphrase Leopold badly). When he wrote, "To keep every cog and wheel is the first precaution of intelligent tinkering," Leopold was referring to the importance of saving large carnivores.[15] Today part of intelligent tinkering, also known as ecological restoration, involves acknowledging that you can't go back, you can only go forward, striving to create healthier ecosystems that preserve essential processes such as predation.[16] So while we can't quite re-create the close relations we may once have had with living things, such as what I experienced in the Great Bear Rainforest, we can envision a world in which we base our relationships with carnivores on respect, rather than fear. A world where we allow them to fulfill their ecological roles as much as possible. A world where we give them room to roam, so that their benefits cascade through whole ecosystems.

Sharing this earth with thriving, healthy carnivores comes down to coexistence. The problem is that *coexistence* means different things to different people. We've seen that to some people, coexistence means keeping carnivore numbers as low as possible short of extermination, in order to produce more moose (*Alces alces*). Via the North American Model of wildlife conservation, this means coexisting with bears, wolves, cougars, and other carnivores on *our* terms, not theirs. Conversely, some define coexistence as protecting every carnivore, completely and always. Realistically, in our fragmented, modern world, coexistence lies somewhere between these two perspectives. And while sometimes it seems like we're very far from achieving such a vision, there are now more large carnivores than there have been in over 100 years in more places than we could have imagined twenty years ago. This gives me hope.

Knowing there are places like Khutze Inlet and Gribbell Island, where bears and humans can just *be*, means that eventually we'll get this right. I'll never find full answers to the questions the bear mother inspired as she looked at me calmly, mother to mother. But she and Scarface taught me that there's no separation here. To me these bears and the other carnivores are walking reminders of why the word *ecology* comes from the Greek work *oikos*—"house." For we're all threads in the same cloth of creation, and we dwell in this Earth household together.[17]

Notes

Introduction

1. Michael Parfitt, "The Hard Ride of Route 93," *National Geographic* 182, no. 6 (1992): 42–69.
2. Aldo Leopold, *A Sand County Almanac: And Sketches Here and There* (New York: Ballantine, 1986), xvii.
3. Ibid., 197.
4. William J. Ripple, James A. Estes, Robert L. Beschta, Christopher C. Wilmers, Euan G. Ritchie, Mark Hebblewhite, Joel Berger, et al., "Status and Ecological Effects of the World's Largest Carnivores," *Science* 343, no. 6167 (2014), 1241484.
5. Aldo Leopold, *Round River*, ed. Luna B. Leopold (Oxford: Oxford University Press, 1993), 146–47.
6. Lionel E. Jackson and Michael C. Wilson, "The Ice-Free Corridor Revisited," *Geotimes* 49, no. 2 (2004): 16–19.
7. David M. Theobold et al., "Connecting Natural Landscapes Using a Landscape Permeability Model to Prioritize Conservation Activities in the United States," *Conservation Letters* 5, no. 2 (2011): 123–33.

Chapter 1: Large Carnivores and Corridor Ecology

1. Tanya Shenk, *Post-Release Monitoring of Lynx (Lynx canadensis) Reintroduced to Colorado* (Fort Collins, CO: Colorado Division of Wildlife, 2011).

2. Paul Beier, "Dispersal of Juvenile Cougars in Fragmented Habitat," *Journal of Wildlife Management* 59, no. 2 (1995): 228–37.

3. Katie Moriarty et al., "Wolverine Confirmation in California after Nearly a Century: Native or Long-Distance Immigrant?" *Northwest Science* 83, no. 2 (2009): 154–62.

4. Charles C. Chester, "Yellowstone to Yukon, North America," in *Climate and Conservation: Landscape and Seascape Science, Planning, and Action,* ed. Jodi A. Hilty, Charles C. Chester, and Molly S. Cross (Washington, DC: Island Press, 2012), 240–52.

5. John L. Weaver, *The Transboundary Flathead: A Critical Landscape for Carnivores in the Rocky Mountains,* Working Paper No. 18 (Bozeman, MT: WCS, 2001), 7–10; Ben Long and the Crown of the Continent Ecosystem Education Consortium, *Crown of the Continent: Profile of a Treasured Landscape* (Dallas, TX: Scott Publishing, 2002).

6. Weaver, ibid.

7. Charles C. Chester, *Conservation across Borders: Biodiversity in an Interdependent World* (Washington, DC: Island Press, 2006), 20–23.

8. Chester, "Yellowstone to Yukon, North America," 240–52.

9. Weaver, *The Transboundary Flathead,* 43–47; also, see the Flathead Wild website: www.flathead.ca/, accessed November 26, 2013.

10. Robert A. Watt, interview by Cristina Eisenberg, August 30, 2012, Waterton Lakes National Park, AB.

11. Grizzlies are protected in Alberta but not British Columbia, as we shall see in chapters 3 and 4. Other species of large carnivores are not protected in Alberta or British Columbia and can be killed by humans legally.

12. Alberta Fish and Wildlife Division, *Management Plan for Wolves in Alberta* (Edmonton, AB: Forestry, Lands, and Wildlife; Fish and Wildlife Division, 1991).

13. Barb Johnston, interview by Cristina Eisenberg, December 4, 2012, Waterton Lakes National Park, AB.

14. Adam T. Ford, Anthony P. Clevenger, and Kathy Rettie, "The Banff Wildlife Crossings Project: An International Public-Private Partnership," in *Safe Passages: Highways, Wildlife, and Habitat Connectivity,* ed. Jon P. Beckmann et al. (Washington, DC: Island Press, 2010), 157–72; Elizabeth Kolbert, *Field Notes from a Catastrophe: Man, Nature, and Climate Change* (New York: Bloomsbury, 2006), 183–89.

15. Chester, "Yellowstone to Yukon, North America," 240–52; Karsten Heuer, *Walking the Big Wild: From Yellowstone to the Yukon on the Grizzly Bear's Trail* (Toronto, ON: McClelland and Stewart, Ltd., 2004), ix–xiv.

16. Reed F. Noss et al., "Conservation Biology and Carnivore Conservation in the Rocky Mountains," *Conservation Biology* 10, no. 4 (1996): 949–63.

17. Wendy Francis, interview by Cristina Eisenberg, September 3, 2012, Banff, AB.

18. Karsten Heuer, interview by Cristina Eisenberg, July 6, 2012, Banff National Park, AB.

19. John Davis, interview by Cristina Eisenberg, November 8, 2013, Seattle, WA.

20. Michael Soulé, interview by Cristina Eisenberg, January 5, 2013, Paonia, CO.

21. Michael E. Soulé and John Terborgh, *Continental Conservation: Scientific Foundations of Regional Reserve Networks* (Washington, DC: Island Press, 1999).

22. Jodi A. Hilty, interview by Cristina Eisenberg, January 30, 2013, Bozeman, MT.

23. Tracy Lee, Michael Quinn, and Danah Duke, "A Local Community Monitors Wildlife along a Major Transportation Corridor," in *Safe Passages*, 277–92; see also: Michael Proctor et al., "Population Fragmentation and Inter-Ecosystem Movements of Grizzly Bears in Western Canada and the Northern United States," *Wildlife Monographs* 180 (2012): 1–46.

24. John L. Weaver, *Conservation Value of Roadless Areas for Vulnerable Fish and Wildlife Species in the Crown of the Continent Ecosystem, Montana*, Working Paper No. 40 (Bozeman, MT: WCS, 2011); John L. Weaver, *Safe Havens, Safe Passages for Vulnerable Fish and Wildlife: Critical Landscapes in the Southern Canadian Rockies, British Columbia and Montana*, Conservation Report No. 6 (Toronto, ON: WCS Canada, 2013); John L. Weaver, *Protecting and Connecting Headwater Havens: Vital Landscapes for Vulnerable Fish and Wildlife Southern Canadian Rockies of Alberta* (Toronto, ON: WCS Canada, 2013).

25. Clayton D. Apps et al., *Carnivores in the Southern Canadian Rockies: Core Areas and Connectivity Across the Crowsnest Highway*, Report No. 3 (Toronto, ON: WCS Canada, 2007); Anthony Clevenger et al., *Highway 3: Transportation Mitigation for Wildlife and Connectivity in the Crown of the Continent Ecosystem* (Bozeman, MT: Western Transportation Institute, 2010).

26. Weaver, *Safe Havens, Safe Passages*, 3–7.

27. John L. Weaver, interview by Cristina Eisenberg, February 21, 2013, St. Ignatius, MT.

28. Proctor et al., "Population Fragmentation and Inter-Ecosystem Movements of Grizzly Bears"; Jodi A. Hilty, interview by Cristina Eisenberg, January 30, 2013, Bozeman, MT.

29. Anthony Clevenger, interview by Cristina Eisenberg, October 9, 2012, Canmore, AB.

30. Anthony P. Clevenger and Adam T. Ford, "Wildlife Crossing Structures, Fencing, and Other Highway Design Considerations," in *Safe Passages*, 17–50; Anthony P. Clevenger and Jack Wierzchowski, "Maintaining and Restoring Connectivity in Landscapes Fragmented by Roads," in *Connectivity Conservation*, ed. Kevin R. Crooks and M. Sanjayan (Cambridge, UK: Cambridge University Press, 2006), 502–35.

31. Anthony P. Clevenger and Nigel Waltho, "Factors Influencing the Effectiveness of Wildlife Underpasses in Banff National Park, Alberta, Canada," *Conservation Biology* 14 (2000): 47–56.

32. Michael A. Sawaya, Anthony P. Clevenger, and Steven T. Kalinowski, "Demographic Connectivity for Ursid Populations at Wildlife Crossing Structures in Banff National Park," *Conservation Biology* 27, no. 4 (2013): 721–30.

33. Anthony Clevenger, interview by Cristina Eisenberg, October 9, 2012, Canmore, AB.

34. Brian Walker and David Salt, *Resilience Thinking: Sustaining Ecosystems and People in a Changing World* (Washington, DC: Island Press, 2006), 12–14.

35. Proctor et al., "Population Fragmentation and Inter-Ecosystem Movements of Grizzly Bears"; John L. Weaver, interview by Cristina Eisenberg, February 21, 2013, St. Ignatius, MT; Walker and Salt, *Resilience Thinking.*

36. Jodi A. Hilty, William Z. Lidicker Jr., and Adina M. Merenlender, *Corridor Ecology: The Science and Practice of Linking Landscapes for Biodiversity Conservation* (Washington, DC: Island Press, 2006), 54–60.

37. John L. Weaver, Paul C. Paquet, and Leonard F. Ruggiero, "Resilience and Conservation of Large Carnivores in the Rocky Mountains," *Conservation Biology* 10, no. 4 (1996): 964–76.

38. Weaver, *Safe Havens, Safe Passages*, 32.

39. Wendy Francis, interview by Cristina Eisenberg, September 3, 2012, Banff, AB.

40. Kevin Van Tighem, interview by Cristina Eisenberg, September 5, 2012, Banff, AB.

Chapter Two: The Ecological Role of Large Carnivores

1. Daniel H. Pletscher et al., "Population Dynamics of a Recolonizing Wolf Population," *Journal of Wildlife Management* 61, no. 2 (1997): 459–65.

2. Robert A. Watt, *Wildlife Reports, Waterton Lakes National Park* (Waterton, AB: Parks Canada, 1980–2009).

3. Robert R. Ream et al., *Population Dynamics and Movements of Recolonizing Wolves in the Glacier National Park Area,* Annual Report (Missoula, MT: University of Montana, 1990).

4. Eliot Fox and Kevin Van Tighem, *The Belly River Wolf Study, Six-Month Interim Report* (Waterton Lakes, AB: Waterton Lakes National Park, 1994); Robert A. Watt, interview by Cristina Eisenberg, August 30, 2012, Waterton Lakes National Park, AB.

5. Joel S. Brown, John W. Laundré, and Mahesh Gurung, "The Ecology of Fear: Optimal Foraging Game Theory and Trophic Interactions," *Journal of Mammalogy* 80, no. 2 (1999): 385–99; John W. Laundré, Lucina Hernandez, and Kelly B. Altendorf,

"Wolves, Elk, and Bison: Re-Establishing the 'Landscape of Fear' in Yellowstone National Park," *Canadian Journal of Zoology* 79 (2001): 1401–9.

6. Robinson Jeffers, "The Bloody Sire," in *The Selected Poetry of Robinson Jeffers* (Palo Alto, CA: Stanford University Press, 2002), 563–64.

7. Robert T. Paine, "A Note on Trophic Complexity and Species Diversity," *The American Naturalist* 103 (1969): 91–93; Robert T. Paine, "Food Webs: Linkage, Interaction Strength, and Community Infrastructure," *Journal of Animal Ecology* 49 (1980): 667–85.

8. Brown et al., "The Ecology of Fear," 486–99.

9. Olaus J. Murie, "Field Notes: Mammals, *Cervus Canadensis*" (Shepherdstown, WV: USFWS Archives, 1926–1954), 417–98; Aldo Leopold, "Deer Irruptions," *Wisconsin Conservation Bulletin* 8 (1943): 3–11.

10. Nelson G. Hairston, Frederick E. Smith, and Lawrence B. Slobodkin, "Community Structure, Population Control, and Competition," *American Naturalist* 94, no. 879 (1960): 421–25.

11. Charles Elton, *Animal Ecology*, 2d ed. (Chicago: University of Chicago Press, 2001), 101–45.

12. Paine, "A Note on Trophic Complexity and Community Stability," 91–93.

13. Michael Soulé et al., "Strongly Interacting Species: Conservation Policy, Management, and Ethics," *Bioscience* 55, no. 2 (2005): 168–76.

14. Wolves most often dig their dens into the root balls of conifers, using the larger roots to create stability. In this particular case, the conifer used was a spruce; however, it was impossible for me to identify the actual species of spruce. In the Crown of the Continent Ecosystem, there are three species of spruce, and all hybridize frequently. Because of this hybridization, identifying spruce by species can only be done via DNA analysis.

15. Cristina Eisenberg, *Complexity of Food Web Interactions in a Large Mammal System* (dissertation, Oregon State University, 2012).

16. William W. Murdoch, "'Community Structure, Population Control, and Competition'—A Critique," *American Naturalist* 100, no. 912 (1966): 219–26.

17. James A. Estes et al., "Trophic Downgrading of Planet Earth," *Science* 33, no. 6040 (2011): 301–6.

18. Charles Darwin, *The World of Charles Darwin*, vol. 16: *The Origin of Species 1876* (New York: NYU Press, 2010).

19. Aldo Leopold, "The Research Program," in *Transactions of the Second North American Wildlife Conference* (1937): 104–7.

20. William J. Ripple et al., "Trophic Cascades among Wolves, Elk, and Aspen on Yellowstone National Park's Northern Range," *Biological Conservation* 102 (2001): 227–34.

21. Cristina Eisenberg, S. Trent Seager, and David E. Hibbs, "Wolf, Elk, and Aspen Food Web Relationships: Context and Complexity," *Forest Ecology and Management* (2013): 70–80.

22. Estes et al., "Trophic Downgrading of Planet Earth"; Oswald J. Schmitz, Peter A. Hamback, and Andrew P. Beckerman, "Trophic Cascades in Terrestrial Systems: A Review of the Effects of Carnivore Removals on Plants," *American Naturalist* 155 (2000): 141–53.

23. Donald Strong, "Are Trophic Cascades All Wet? Differentiation and Donor Control in Speciose Systems," *Ecology* 73 (1992): 745–54; Mark Hebblewhite et al., "Human Activity Mediates a Trophic Cascade Caused by Wolves," *Ecology* 86 (2005): 2135–44.

24. Matthew Kauffman, Jedediah F. Brodie, and Erik S. Jules, "Are Wolves Saving Yellowstone's Aspen? A Landscape-Level Test of a Behaviorally Mediated Trophic Cascade," *Ecology* 91 (2010), 2742–55.

25. The Waterton and Saint Mary Valleys in Glacier National Park have had fires in recent times but none in the aspen communities I studied during the years of my research (2007–11). There were attempts to set some prescribed burns in my Waterton Valley study plots. With one exception these fizzled out. The one fire that did take off resulted in a postdoctoral research project for me that is still underway as I write this, but that was not part of my doctoral research; see: Eisenberg, *Complexity of Food Web Interactions*, 134–80.

26. Eisenberg, *Complexity of Food Web Interactions*, 87–133.

27. Tony K. Ruth and Kerry Murphy, "Cougar–Prey Relationships," in *Cougar Ecology and Conservation*, ed. Maurice Hornocker and Sharon Negri (Chicago: University of Chicago Press, 2010), 138–62; cougars have been found to kill adult elk regularly in northeast Oregon, but frequent cougar predation on adult elk has not been reported in other places. Cougar predation on elk is not fully understood (e.g., under what circumstances it does or does not take place), despite many studies on cougar predation.

28. Charles J. Krebs et al., "What Drives the Ten-Year Cycle of Snowshoe Hares?" *BioScience* 51, no. 1 (2001): 25–35.

29. Aldo Leopold, *Round River*, ed. Luna B. Leopold (Oxford: Oxford University Press, 1993), 147.

30. Soulé et al., "Ecological Effectiveness"; Cristina Eisenberg et al., "Wolf, Elk, and Aspen Food Web Relationships," 70–80.

31. Edward O. Wilson, *The Diversity of Life* (Cambridge, MA: Belknap Press of Harvard University Press, 1992); Elizabeth Kolbert, *Field Notes from a Catastrophe: Man, Nature, and Climate Change* (New York: Bloomsbury, 2006), 183–89.

32. William Obadiah Pruitt and Leonid M. Baskin, *Boreal Forest of Canada and Russia* (Sofia, Bulgaria: Pensoft Publishers, 2004), 1–167.

33. W. D. Billings and H. A. Mooney, "The Ecology of Arctic and Alpine Plants," *Biological Review* 43 (1968): 481–529.

34. Robert L. Beschta and William J. Ripple, "Rapid Assessment of Riparian Cottonwood Recruitment: Middle Fork John Day River, Northeastern Oregon," *Ecological Restoration* 23, no. 3 (2005): 150–56.

35. Charles C. Schwartz, Albert W. Franzmann, Richard E. McCabe, *Ecology and Management of the North American Moose* (Boulder, CO: University Press of Colorado, 2007).

36. Charles Krebs et al., "Terrestrial Trophic Dynamics in the Canadian Arctic," *Canadian Journal of Zoology* 81 (2003): 827–43.

Chapter Three: Crossings

1. Robert B. Keiter and Harvey Locke, "Law and Large Carnivore Conservation in the Rocky Mountains of the U.S. and Canada," *Conservation Biology* 10, no 4 (1996): 1003–12.

2. Charles Chester, *Conservation across Borders: Biodiversity in an Independent World* (Washington, DC: Island Press, 2006), 14–52.

3. Richard West Sellars, *Preserving Nature in the National Parks: A History* (New Haven, CT: Yale University Press, 1997), 3–5; Raul Valdez et al., "Wildlife Conservation and Management in Mexico," *Wildlife Society Bulletin* 34, no. 2 (2006): 270–82.

4. Peyton Doub, *The Endangered Species Act: History, Implementation, Successes, and Controversies* (Boca Raton, FL: Taylor and Francis, 2013), 3–14.

5. Aldo Leopold, *A Sand County Almanac: And Sketches Here and There* (New York: Ballantine, 1986); Rachel Carson, *Silent Spring* (New York: Houghton Mifflin, 1962).

6. Kieran F. Suckling and Martin Taylor, "Critical Habitat and Recovery," in Dale D. Goble, J. Michael Scott, and Frank Davis, eds., *The Endangered Species Act at Thirty*, vol. 1: *Renewing the Conservation Promise* (Washington, DC: Island Press, 2006), 75–89.

7. 16 U.S.C. §1532(6); 16 U.S.C. §1532(20); http://ecos.fws.gov/tess_public/pub/boxScore.jsp, accessed March 1, 2013.

8. Douglas Honnold, Managing Attorney, EarthJustice Legal Defense Fund, interview by Cristina Eisenberg, January 22, 2013, San Francisco, CA; Holly Doremus, "Lessons Learned," *The Endangered Species Act at Thirty*, 195–207.

9. D. Noah Greenwald, Kieran F. Suckling, and Martin Taylor, "The Listing Record," *The Endangered Species Act at Thirty*, 51–67.

10. Doub, *The Endangered Species Act*, 208–32.

11. 16 U.S.C. §1532 (3)(6).

12. Douglas Honnold, Managing Attorney, EarthJustice Legal Defense Fund, interview by Cristina Eisenberg, January 22, 2013, San Francisco, CA.

13. Ibid.

14. Ibid.

15. Doub, *The Endangered Species Act*, 52–57.

16. USFWS, *The Reintroduction of Gray Wolves to Yellowstone National Park and Central Idaho: Final Environmental Impact Statement* (Denver, CO: USFWS, 1994).

17. USFWS, "Final Rule to Identify the Northern Rocky Mountain Population of Gray Wolf as a Distinct Population Segment and to Revise the List of Endangered and Threatened Wildlife," CFR Part 17, *Federal Register* 74, no. 62 (April 2, 2013), 15123.

18. Doub, *The Endangered Species Act*, 174–76; Keiter and Locke, "Law and Large Carnivore Conservation in the Rocky Mountains of the US and Canada," 1003–12.

19. Jamie Benidickson, *Environmental Law*, 3d ed. (Ottawa: Irwin Law, 2009), 316–19.

20. S.C. 2002, c. 29 § 2(1).

21. Benidickson, *Environmental Law*, 100–17.

22. Raul Valdez et al., "Wildlife Conservation and Management in Mexico," 270–82; Environmental Law Institute, *Decentralization of Environmental Protection in Mexico: An Overview of State and Local Laws and Institutions* (Washington, DC: Environmental Law Institute, 1996).

23. Eric T. Freyfogle and Dale D. Goble, *Wildlife Law: A Primer* (Washington, DC: Island Press, 2009), 164–83.

24. Benidickson, *Environmental Law*, 41–44, 218.

25. Rachel Rose Starks and Adrian Quijada-Mascarenas, "A Convergence of Borders: Indigenous Peoples and Environmental Conservation at the US-Mexico Border," in *Conservation of Shared Environments: Learning from the United States and Mexico*, ed. Laura Lopez-Hoffman et al. (Tucson: University of Arizona Press, 2009), 54–70.

26. Reed F. Noss, "From Endangered Species to Biodiversity," in Kathryn A. Kohm, ed., *Balancing on the Brink of Extinction: The Endangered Species Act and Lessons for the Future* (Washington, DC: Island Press, 1991), 236.

27. Randall C. Archibold, "Border Plan Will Address Harm Done at Fence Site," *New York Times*, January 16, 2009, www.nytimes.com/2009/01/17/us/17border.html?ref=borderfenceusmexico, accessed December 15, 2012.

28. Richard R. Frankham, Jonathan D. Ballou, David A. Briscoe, *Introduction to Conservation Genetics* (Cambridge, UK: Cambridge University Press, 2010).

29. Brian P. Segee and Ana Cordova, "A Fence Runs through It," in *Conservation of Shared Environments*, 241–56.

30. Brad McRae et al., "Habitat Barriers Limit Gene Flow and Illuminate Historical Events in a Wide-Ranging Carnivore, the American Puma," *Molecular Ecology* 14 (2005): 1965–77.

31. Melanie Culver et al., "Connecting Wildlife Habitats across the US-Mexico Border," in *Conservation of Shared Environments*, 83–99; Ana Cordova and Carlos A. de la Parva, "Transboundary Conservation between the United States and Mexico: New Institutions or a New Collaboration?" in *Conservation of Shared Environments*, 279–92.

32. Aldo Leopold, "Foreword" (original foreword to *A Sand County Almanac*), Aldo Leopold Archives, University of Wisconsin, unpublished manuscripts, 10-6, Box 16.

33. Wallace Stegner, *The Sound of Mountain Water* (New York: Penguin Books, 1969), 153.

Chapter 4: Grizzly Bear

1. Timothy Rawson, *Changing Tracks: Predators and Politics in Mt. McKinley National Park* (Fairbanks, AK: University of Alaska Press, 2001).

2. Adolph Murie, *The Grizzlies of Mount McKinley* (Seattle: University of Washington Press, 1981); Rawson, *Changing Tracks*.

3. Bruce N. McLellan, Chris Servheen, and Djuro Huber (IUCN SSC Bear Specialist Group), "*Ursus arctos*," in *IUCN Red List of Threatened Species*, v. 2012.2 (2008), www.iucnredlist.org, accessed March 18, 2013; Paul Schullery, *Lewis and Clark among the Grizzlies* (Helena: Falcon Guides, 2002), 41–54; Robert H. Busch, *The Grizzly Almanac* (New York: Lyons Press, 2000), 9–14.

4. Kevin Van Tighem, *Bears: Without Fear* (Toronto: Rocky Mountain Books, 2013), 64.

5. Pierpont Morgan Library Manuscript MS M871 (Plinus Segundus Gaius, between 830 and 840).

6. John Craighead, Jay Sumner, and John Alexander Mitchell, *The Grizzly Bears of Yellowstone: Their Ecology in the Yellowstone Ecosystem* (Washington, DC: Island Press, 1995), 383–84; Van Tighem, *Bears: Without Fear*, 176.

7. Craighead et al., *The Grizzly Bears of Yellowstone*, 155–92; John L. Weaver, Paul C. Paquet, and Leonard F. Ruggiero, "Resilience and Conservation of Large Carnivores in the Rocky Mountains," *Conservation Biology* 10, no. 4 (1996): 964–76.

8. David J. Mattson, "Use of Ungulates by Yellowstone Grizzly Bears (*Ursus arctos*)," *Biological Conservation* 181 (1997): 161–77; John S. Waller and Richard D. Mace, "Grizzly Bear Habitat Selection in the Swan Mountains, Montana," *Journal of Wildlife Management* 61, no. 4 (1997): 1032–9.

9. George V. Hilderbrand et al., "The Importance of Meat, Particularly Salmon, to Body Size, Population Productivity, and Conservation of North American Brown Bears," *Canadian Journal of Zoology* 77 (1999): 132–38; David J. Mattson, Bonnie M. Blanchard, and Richard R. Knight, "Food Habits of Yellowstone Grizzly Bears," *Journal of Applied Ecology* 34 (1991): 926–40.

10. David J. Mattson, Bonnie M. Blanchard, and Richard R. Knight, "Yellowstone Grizzly Bear Mortality, Human Habituation, and Whitebark Pine Seed Crops," *Journal of Wildlife Management* 56, no. 3 (1992): 432–42; Katherine C. Kendall and Robert E. Keane, "Whitebark Pine Decline: Infection, Mortality, and Population Trends," in *Whitebark Pine Communities: Ecology and Restoration*, ed. Diana F. Tomback, Stephen F. Arno, and Robert E. Keane (Washington, DC: Island Press, 1996), 221–42.

11. Van Tighem, *Bears: Without Fear*, 69–70.

12. Richard D. Mace et al., "Landscape Evaluation of Grizzly Bear Habitat in Western Montana," *Conservation Biology* 13, no. 2 (1999): 367–77; John Waller and Christopher Servheen, "Effects of Transportation Infrastructure on Grizzly Bears in Northwestern Montana," *Journal of Wildlife Management* 69 (2005): 985–1000.

13. Chris Wilmers et al., "Resource Dispersion and Consumer Dominance: Scavenging and Wolf- and Hunter-Killed Carcasses in Greater Yellowstone USA," *Ecology Letters* 6 (2003): 996–1003; Rolf O. Peterson, "Temporal and Spatial Aspects of Predator-Prey Dynamics," *Alces* 39 (2003): 215–32.

14. Sandra E. Tardiff and Jack A. Stanford, "Grizzly Bear Digging: Effects on Subalpine Meadow Plants in Relation to Mineral Nitrogen Availability," *Ecology* 79, no. 7 (1998): 2219–28.

15. Gordon W. Holtgrieve, Daniel E. Schindler, and Peter K. Jewett, "Large Predators and Biogeochemical Hotspots: Brown Bear (*Ursus arctos*) Predation on Salmon Alters Nitrogen Cycling in Riparian Soils," *Ecological Research* 24, no. 5 (2009): 1125–35.

16. Tracy I. Storer and Lloyd P. Tevis Jr., *California Grizzly* (Berkeley, CA: University of California Press, 1996) ; Busch, *The Grizzly Almanac*, 14–15; David J. Mattson, and Troy Merrill, "Extirpations of Grizzly Bears in the Contiguous United States, 1850–2000," *Conservation Biology* 16, no. 4 (2002): 1123–36.

17. Mattson and Merrill, "Extirpations of Grizzly Bears in the Contiguous United States," 1123–36.

18. USFWS, *Grizzly Bear Recovery Plan* (Missoula, MT: USFWS, 1993); USFWS, "Grizzly Bear Recovery," www.fws.gov/mountain-prairie/species/mammals/grizzly/, accessed March 27, 2013.

19. Michael Lang, *Bears: Tracks through Time* (Banff, AB: The Whyte Museum of the Canadian Rockies, 2010), 80–87.

20. Jeff Gailus, *The Grizzly Manifesto: In Defense of the Great Bear* (Vancouver, BC: Rocky Mountain Books, 2010); Alberta SRD, "Wildlife Species Status," www .srd.alberta.ca/FishWildlife/SpeciesAtRisk/GeneralStatusOfAlbertaWildSpecies /GeneralStatusOfAlbertaWildSpecies2010/SearchForWildSpeciesStatus.aspx, accessed September 28, 2012; Steve Michel, "Living with Wildlife," www.youtube .com/watch?v=2M5M2jR2VaQ, accessed November 21, 2013.

21. Kerry Gunther, interview by Cristina Eisenberg, September 18, 2012, Mammoth, WY.

22. Miguel Llanos, "Grizzly Mauls Man to Death at Denali; Bear Shot," *NBC News*, August 25, 2012.

23. Stephen Herrero, interview by Cristina Eisenberg, December 5, 2012, Calgary, AB.

24. Steven Michel, interview by Cristina Eisenberg, October 9, 2012, Banff National Park, AB; Herrero, *Bear Attacks: Their Causes and Avoidance* (Guilford, CT: Lyons Press, 2002), 41.

25. Herrero, *Bear Attacks*; Van Tighem, *Bears: Without Fear*.

26. Charlie Russell, interview by Cristina Eisenberg, August 21, 2012, The Hawks Nest, Waterton, AB.

27. Steven Michel, interview by Cristina Eisenberg, October 9, 2012, Banff National Park, AB; Kerry Gunther, interview by Cristina Eisenberg, September 18, 2012, Mammoth, WY; John Waller, Glacier National Park ecologist, e-mail dated April 5, 2013.

28. Glacier National Park Archives, case incident record, August 1, 1976.

29. Details on Timothy Treadwell's work with bears and the events that led to his death are provided in the Nick Jans book, *The Grizzly Maze* (New York: Plume, 2005); details about Treadwell's disregard of advice was further provided by Charlie Russell, interview by Cristina Eisenberg, August 21, 2012, The Hawks Nest, Waterton, AB.

30. Craig Boddington, *Fair Chase in North America* (Missoula, MT: The Boone and Crockett Club, 2004).

31. Cathy Ellis, "Famous Grizzly Sow Defies Odds," *Rocky Mountain Outlook*, July 28, 2011.

32. Van Tighem, *Bears: Without Fear*, 69–70.

33. Steven Michel, interview by Cristina Eisenberg, October 9, 2012, Banff National Park, AB.

34. Ibid.

35. Katherine C. Kendall et al., "Demography and Genetic Structure of a Recovering Grizzly Bear Population," *Journal of Wildlife Management* 73, no. 1 (2009): 3–17; John Waller, Glacier National Park ecologist, e-mail dated April 5, 2013.

36. John Waller, Glacier National Park ecologist, e-mail dated April 5, 2013.

37. USFWS, *Final Conservation Strategy for the Grizzly Bear in the Greater Yellowstone Area* (Missoula, MT: USFWS, 2007); USFWS, *Grizzly Bear Recovery Plan, Supplement: Habitat-Based Recovery Criteria for the Yellowstone Ecosystem* (Missoula, MT: USFWS, 2007); USFWS, *Grizzly Bear Recovery Plan, Supplement: Revised Demographic Recovery Criteria for the Yellowstone Ecosystem* (Missoula, MT: USFWS, 2007).

38. Douglas Honnold, interview by Cristina Eisenberg, January 22, 2013, San Francisco, CA.

39. Kerry Gunther, interview by Cristina Eisenberg, September 18, 2012, Mammoth, WY.

40. USFWS, *Grizzly Bear Recovery Plan, Supplement.*

41. David J. Mattson, "Sustainable Grizzly Bear Mortality Calculated from Counts of Females with Cubs-of-the-Year: An Evaluation," *Biological Conservation* 81 (1997): 103–11; USFWS, *Grizzly Bear Recovery Plan, Supplement: Proposed Application Protocol for Yellowstone Grizzly Bear Demographic Recovery Criteria* (Missoula, MT: USFWS, 2013).

42. Daniel F. Doak and Kerry Cutler, "Re-evaluating Evidence for Past Population Trends and Predicted Dynamics of Yellowstone Grizzly Bears," *Conservation Letters* (2013) DOI: 10.1111/conl.12048.

43. Christopher Servheen, interview by Cristina Eisenberg, February 20, 2013, Missoula, MT; Michael Proctor et al., "Population Fragmentation and Inter-Ecosystem Movements of Grizzly Bears in Western Canada and the Northern United States," *Wildlife Monographs* 180 (2012): 1–46.

44. Christopher Servheen, interview by Cristina Eisenberg, February 20, 2013, Missoula, MT; Kerry A. Gunther et al., "Grizzly Bear–Human Conflicts in the Greater Yellowstone Ecosystem, 1992–2000," *Ursus* 15 (2004): 10–22.

45. Susan Clark, Murray B. Rutherford, and Denise Casey, eds., *Coexisting with Large Carnivores: Lessons from the Greater Yellowstone* (Washington, DC: Island Press, 2005).

46. Peter H. Kahn Jr. and Patricia H. Hasbach, eds., *The Rediscovery of the Wild* (Cambridge, MA: MIT Press, 2013); Peter H. Kahn, "Cohabitating with the Wild," *Ecopsychology*, March 2009: 38–46.

Chapter 5: Wolf

1. L. David Mech et al., *The Wolves of Denali* (Minneapolis, MN: University of Minnesota Press, 1998), 159–74.

2. Tom Meier, interview by Cristina Eisenberg, July 12, 2012, Denali, AK.

3. NPS, *Revisiting Leopold: Resource Stewardship in the National Parks* (Washington, DC: National Park Service, 2012); Douglas W. Smith, interview by Cristina Eisenberg, January 30, 2013, Bozeman, MT.

4. Timothy Rawson, *Changing Tracks: Predators and Politics in Mt. McKinley National Park* (Fairbanks, AK: University of Alaska Press, 2001), 206–24.

5. Adolph Murie, *The Wolves of Mount McKinley* (Washington, DC: National Park Service, 1944); Rawson, *Changing Tracks*, 237.

6. Mech et al., *The Wolves of Denali*; Gordon Haber and Marybeth Holleman, *Among Wolves* (Fairbanks, AK: University of Alaska Press, 2013); Vic Van Vallenberghe, *In the Company of Moose* (Mechanicsburg, PA: Stackpole Books, 2004).

7. Ronald Nowak, "Wolf Evolution and Taxonomy," in *Wolves: Behavior, Ecology, and Conservation*, ed. L. David Mech and Luigi Boitani (Chicago: University of Chicago Press, 2003), 239–50; Garry Marvin, *Wolf* (London: Reaktion Books, 2012), 13–14.

8. Stuart M. Chambers et al., "An Account of the Taxonomy of North American Wolves from Morphological and Genetic Analyses," *North American Fauna* 77 (2012): 1–67.

9. L. David Mech, *The Wolf* (Minneapolis, MN: University of Minnesota Press, 1970), 11–12; Rolf O. Peterson and Paolo Ciucci, "The Wolf as a Carnivore," in *Wolves: Behavior, Ecology, and Conservation*, 112–13; Kevin Van Tighem, *The Homeward Wolf* (Toronto: Rocky Mountain Books, 2013), 14.

10. Murie, *The Wolves of Mount McKinley*; Haber and Holleman, *Among Wolves*, 31–58; Jane M. Packard, "Wolf Behavior: Reproductive, Social, and Intelligent," in *Wolves: Behavior, Ecology, and Conservation*, 53–55.

11. Diane K. Boyd and Daniel H. Pletscher, "Characteristics of Dispersal in a Colonizing Wolf Population in the Central Rocky Mountains," *Journal of Wildlife Management* 63 (1999): 1094–1108.

12. Mech and Boitani, "Wolf Social Ecology," 31; Packard, "Wolf Behavior," 46; Douglas W. Smith, e-mail to author dated January 9, 2014.

13. Douglas W. Smith and Michael K. Phillips, "Northern Rocky Mountain Wolf (*Canis lupus nubilus*)," in *Endangered Animals: A Reference Guide to Conflicting Issues*, ed. Richard P. Reading and Brian Miller (Westport, CT: Greenwood Press, 2000); Kyran E. Kunkel et al., "Factors Affecting Foraging Behavior of Wolves in and near Glacier National Park, Montana," *Journal of Wildlife Management* 68 (2004): 167–78.

14. Vanessa Renwick, *Hunting Requires Optimism* (multi-media installation), www.odoka.org/the_work/, accessed April 18, 2013.

15. Douglas W. Smith, e-mail to author dated January 9, 2014.

16. Olaus J. Murie, "Field Notes: Mammals, *Cervus Canadensis*" (Shepherdstown, WV: USFWS Archives, 1926–1954), 417–98; Aldo Leopold, *A Sand County Almanac: And Sketches Here and There* (New York: Ballantine, 1986), 135.

17. Nelson G. Hairston, Frederick E. Smith, and Lawrence B. Slobodkin, "Community Structure, Population Control, and Competition," *American Naturalist* 94, no. 879 (1960): 421–25.

18. Rick McIntyre, ed., *War Against the Wolf: America's Campaign to Exterminate the Wolf* (Stillwater, MN: Voyageur Press, 1995).

19. USFWS, *The Reintroduction of Gray Wolves to Yellowstone National Park and Central Idaho: Final Environmental Impact Statement* (Denver, CO: USFWS, 1994); USFWS, *Reintroduction of the Mexican Gray Wolf within Its Historic Range in the Southwestern US, Final Environmental Impact Statement* (Washington, DC: USFWS, 1996).

20. Douglas W. Smith, e-mail to author dated January 9, 2014.

21. Edward Bangs et al., "Managing Wolf–Human Conflict in the Northwestern United States," in *People and Wildlife: Conflict or Coexistence*, ed. Rosie Woodroffe, Simon Thirgood, and Alan Rabinowitz (Cambridge, UK: Cambridge University Press, 2005), 340–56; USFWS, "U.S. Fish and Wildlife Service, State Agencies Release 2012 Annual Report for Northern Rocky Mountain Wolf Population" (media release), April 13, 2013.

22. David Parsons, interview by Cristina Eisenberg, September 25, 2012, Albuquerque, NM.

23. USFWS, "Mexican Wolf Blue Range Reintroduction Project Statistics," www.fws.gov/southwest/es/mexicanwolf/MWPS.cfm, accessed April 4, 2013; Paul Paquet et al., *Mexican Wolf Recovery: Three-Year Program Review* (Washington, DC: USFWS, 2001).

24. Mark E. McNay, "Wolf–Human Interactions in Alaska and Canada: A Review of the Case History," *Wildlife Society Bulletin* 30 (2002): 831–43.

25. Douglas W. Smith, e-mail dated April 22, 2013.

26. USFWS, "Final Rule to Identify the Northern Rocky Mountain Population of Gray Wolf as a Distinct Population Segment and to Revise the List of Endangered and Threatened Wildlife," CFR Part 17, *Federal Register* 74, no. 62 (April 2, 2013): 15123; John Vucetich, interview by Cristina Eisenberg, April 16, 2013, Corvallis, OR.

27. John Horning, interview by Cristina Eisenberg, September 23, 2012, Santa Fe, NM; David Parsons, interview by Cristina Eisenberg, September 25, 2012, Albuquerque, NM.

28. USFWS, "Mexican Wolf Blue Range Reintroduction Project Statistics," www.fws.gov/southwest/es/mexicanwolf/MWPS.cfm, accessed April 4, 2013.

29. David Parsons, interview by Cristina Eisenberg, September 25, 2012, Albuquerque, NM.

30. Amaroq Weiss, interview by Cristina Eisenberg on April 23, 2013, Petaluma, CA.

31. USFWS, "Removing the Gray Wolf (*Canis lupus*) from the List of Endangered and Threatened Wildlife and Maintain Protections for the Mexican Wolf (*Canis l. baileyi*) by Listing It as Endangered," CFR Part 17, *Federal Register* 78, no. 91 (October 2, 2013): 60813–15.

32. The use of terms such as *take* to mean killing, *liberal* to refer to widespread killing, or *harvest* to mean killing an animal takes us into environmental ethical terrain. Some believe that because animals are living, sentient beings, it is unethical to use such language to refer to killing them. In this book, when referring to laws and public policies I adhere to the language used in these documents, but in my review and analysis I try to use common language (e.g., *killing* or *hunting* instead of *take*) for clarity and to acknowledge the ethical questions raised (or obscured) by terminology.

33. Alaska Department of Fish and Game, "Species Account: Wolf (*Canis lupus*)," www.adfg.alaska.gov/index.cfm?adfg=wolf.main, accessed April 4, 2013; ADFG, *Intensive Management Protocol* (Anchorage, AK: ADFG, 2011).

34. Bob Hayes, *Wolves of the Yukon* (Smithers, BC: Bob Hayes, 2010), 248–59.

35. John Vucetich and Michael Nelson, "What Are Sixty Warblers Worth? Killing in the Name of Conservation," *Oikos* 116 (2007): 1267–78.

36. Rick McIntyre, interview by Cristina Eisenberg, September 16, 2012, Silver Gate, MT; Norm Bishop, interview by Cristina Eisenberg, November 25, 2012, Bozeman, MT.

37. P. J. White, Kelly M. Proffitt, and Thomas O. Lemke, "Changes in Elk Distribution and Group Sizes after Wolf Restoration, *American Midland Naturalist* 167 (2012): 174–87.

38. Douglas W. Smith et al., *Yellowstone Wolf Project: Annual Report, 2011* (Yellowstone National Park, WY: NPS, Yellowstone Center for Resources, 2012).

39. Scott Creel and Jay J. Rotella, "Meta-Analysis of Relationships Between Human Offtake, Total Mortality, and Population Dynamics of Gray Wolves (*Canis lupus*)," *PloS ONE* 5, no. 9 (2010): 1–7, e12918.

40. Layne Adams et al. "Population Characteristics and Harvest Dynamics of Wolves in the Central Brooks Range," *Wildlife Monographs* 170 (2008): 1–25; Todd Fuller, L. David Mech, and Jean F. Cochrane, "Wolf Population Dynamics," in *Wolves: Behavior, Ecology, and Conservation*, 183–84.

41. Creel and Rotella, "Meta-Analysis of Relationships Between Human Offtake, Total Mortality, and Population dynamics of Gray Wolves"; Dennis L. Murray et al., "Death from Anthropogenic Causes Is Partially Compensatory in Recovering Wolf Population," *Biological Conservation* 143 (2010): 2514–24.

42. Douglas W. Smith, interview by Cristina Eisenberg, January 30, 2013, Bozeman, MT.

43. Haber and Holleman, *Among Wolves.*

44. Rick McIntyre, interview by Cristina Eisenberg, September 16, 2012, Silver Gate, MT; Douglas W. Smith, e-mail dated April 22, 2013.

45. Bridget Borg, interview by Cristina Eisenberg, July 18, 2013, Denali, AK; www.nps .gov/dena/naturescience/wolfviewing.htm, accessed December 3, 2013.

46. Jane Goodall, "The Last Wolves?," *Washington Post*, January 5, 2014, www.washing tonpost.com/opinions/the-last-wolves/2014/01/03/621c5d26-71bb-11e3-8def-a 33011492df2_gallery.html, accessed January 19, 2014.

47. Van Tighem, *The Homeward Wolf.*

48. Carter Niemeyer, *Wolfer* (Boise, ID: Bottlefly Press, 2010).

49. Louise Liebenberg, "Managing Our Predators on Our Northern Alberta Ranch," *Shepherd* 56, no. 9: 26–27.

50. Steve Clevidence interview by Cristina Eisenberg, February 19, 2013, Stevensville, MT.

51. Olivier LaRoque, "The Wolves Are Back" (dissertation, McGill University, 2013).

Chapter 6: Wolverine

1. Douglas H. Chadwick, *The Wolverine Way* (Ventura, CA: Patagonia Books, 2010), 47, 235–36.

2. Jeffrey P. Copeland and Richard E. Yates, *Wolverine Population Assessment in Glacier National Park: Comprehensive Summary Update* (Missoula, MT: USDA Forest Service, Rocky Mountain Research Station, 2008).

3. Eric Tomasik and Joseph A. Cook, "Mitochondrial Phylogeography and Conservation Genetics of Wolverine (*Gulo gulo*) of Northwestern North America," *Journal of Mammalogy* 86, no. 2 (2005): 386–96.

4. W. Christopher Wozencraft, "Order Carnivora," in *Mammal Species of the World: A Taxonomic and Geographic Reference*, ed. Don E. Wilson and DeAnn M. Reeder (Baltimore, MD: Johns Hopkins University Press, 2005), 601–19.

5. Vivian Banci, "Wolverine," in *The Scientific Basis for Conserving Forest Carnivores*, General Technical Report RM-254, ed. Leonard F. Rugiero et al. (Fort Collins, CO: USDA Forest Service, 1994), 99–127.

6. Audrey J. Magoun et al., "Modeling Wolverine Occurrence Using Aerial Surveys of Tracks in Snow," *Journal of Wildlife Management* 71, no. 7 (2007): 2221–29.

7. Robert A. Long et al., *Noninvasive Survey Methods for Carnivores* (Washington, DC: Island Press, 2008).

8. Robert Inman et al., *Greater Yellowstone Wolverine Program*, Progress Report (Bozeman, MT: WCS, 2009).

9. Jeffrey P. Copeland, "Seasonal Habitat Associations of the Wolverine of Central Idaho," *Journal of Wildlife Management* 71, no. 7 (2007): 2201–12; Evelyn L. Bull, Keith B. Aubry, and Barbara C. Wales, "Effects of Disturbance on Forest Carnivores of Conservation Concern in Eastern Oregon and Washington," *Northwest Science* 75 (2001): 180–84; John Krebs, Eric C. Lofroth, and Ian Parfitt, "Multiscale Habitat Use by Wolverines in British Columbia," *Journal of Wildlife Management* 71, no. 7

(2007): 2180–92; Jeffrey P. Copeland et al., "The Bioclimatic Envelope of the Wolverine (*Gulo gulo*): Do Climatic Constraints Limit Its Geographic Distribution?" *Canadian Journal of Zoology* 88 (2011): 233–46.

10. Inman et al., *Greater Yellowstone Wolverine Program*, Progress Report 2009; Copeland and Yates, *Wolverine Population Assessment in Glacier National Park*; Jens Person, "Female Wolverine (*Gulo gulo*) Reproduction: Reproductive Costs and Winter Food Availability," *Canadian Journal of Zoology* 83 (2005): 1453–59; John L. Weaver, Paul, C. Paquet, and Leonard F. Ruggiero, "Resilience and Conservation of Large Carnivores in the Rocky Mountains," *Conservation Biology* 10, no. 4 (1996): 964–76.

11. Audrey J. Magoun and Jeffrey P. Copeland, "Characteristics of Wolverine Reproductive Den Sites," *Journal of Wildlife Management* 62, no. 4 (1998): 1313–20.

12. Copeland and Yates, *Wolverine Population Assessment in Glacier National Park*.

13. Robert M. Inman et al., "Spatial Ecology of Wolverines at the Southern Periphery of Distribution," *Journal of Wildlife Management* 76, no. 4 (2011): 778–92.

14. William F. Wood et al., "Potential Semiochemicals in Urine from Free-Ranging Wolverines (*Gulu gulo Pallas*, 1780), *Biochemical Systematics and Ecology* 37, no. 5 (2009): 574–78.

15. Robert Inman et al., "Wolverine Makes Extensive Movements in the Greater Yellowstone Ecosystem," *Northwest Science Notes* 78, no. 3 (2004): 261–66; Katie Moriarty et al., "Wolverine Confirmation in California after Nearly a Century: Native or Long-Distance Immigrant?" *Northwest Science* 83, no. 2 (2009): 154–62; Inman et al., *Greater Yellowstone Wolverine Program*, Progress Report, 2009.

16. Eric C. Lofroth et al., "Food Habits of Wolverine *Gulo gulo* in Montane Ecosystems of British Columbia," *Wildlife Biology* 13 (2007): 31–37.

17. John Krebs et al., "Synthesis of Survival Rates and Causes of Mortality in North American Wolverines," *Journal of Wildlife Management* 68, no. 3 (2004): 493–502.

18. John Lee and Allen Niptanatiak, *Ecology of the Wolverine on the Central Arctic Barrens: Progress Report, Spring 1993* (Yellowknife, NT: Department of Renewable Resources, 1993); Robert M. Inman et al., "Wolverine Space Use in Greater Yellowstone," in *Wildlife Conservation Society, Greater Yellowstone Wolverine Program, Cumulative Report* (Bozeman, MT: WCS, 2007).

19. Maurice G. Hornocker and Howard S. Hash, "Ecology of the Wolverine in Northwest Montana," *Canadian Journal of Zoology* 59 (1981): 1286–1301.

20. James D. Keyser, *Indian Rock Art of the Columbia Plateau* (Seattle and London: University of Washington Press, 1992).

21. Environment BC, "Furbearer Management Guidelines," www.env.gov.bc.ca/fw /wildlife/hunting/regulations/1214/docs/trapping_Section.pdf, accessed May 5, 2013.

22. Eric C. Lofroth and Peter K. Ott, "Assessment of the Sustainability of Wolverine Harvest in British Columbia, Canada," *Journal of Wildlife Management* 71, no. 7 (2007): 2193–2200; Robert M. Inman et al., "Wolverine Harvest in Montana: Survival Rates and Spatial Considerations for Harvest Management," in *Greater Yellowstone Wolverine Program, Cumulative Report, May 2007* (Ennis, MT: Wildlife Conservation Society, 2007), 85–97.

23. TWS, "Final Position Statement, Traps, Trapping, and Furbearer Management, March 2010," http://joomla.wildlife.org/documents/positionstatements/09 -Trapping.pdf, accessed May 11, 2013; FWP, "Trapping and Furbearer Management in Montana," http://fwp.mt.gov/hunting/trapping/, accessed April 28, 2013.

24. Douglas Chadwick, interview by Cristina Eisenberg, Whitefish Montana, September 15, 2014.

25. USFWS, "12-Month Finding on a Petition to List the North American Wolverine as Endangered or Threatened," 50 CFR Part 17, Federal Register 72, no. 107 (June 5, 2007): 31048–9; 16 U.S.C §1532 (3)(20); Doub, *The Endangered Species Act*, 208–32.

26. Inman et al., "Spatial Ecology of Wolverines." While a DPS is defined by Congress as a population of a species that is endangered or threatened, a DPS is tied to a specific geographic location. In order to be recognized, a DPS must be formally defined by USFWS. This definition is posted as a "rule" in the *Federal Register*. Until that has been done, any existing population of an endangered or threatened species does not have legal standing.

27. USFWS, "Initiation of Status Review of the North American Wolverine in the Contiguous United States," 50 CFR Part 17, *Federal Register* 75, no. 72 (April 15, 2010): 19591–92.

28. CDOW, Draft Plan to Reintroduce Wolverine (*Gulo gulo*) to Colorado (Denver, CO: CDOW, 2010); USFWS, "Threatened Status for the Distinct Population Segment of the North American Wolverine," CFR Part 17, *Federal Register* 78, no. 23 (February 4, 2013): 7864–90.

29. Eric C. Lofroth and John Krebs, "The Abundance and Distribution of Wolverines in British Columbia, Canada," *Journal of Wildlife Management* 71, no. 7 (2007): 2159–69; Environment BC, "Furbearer Management Guidelines"; Alberta SRD, "Species Summary: Wolverine (*Gulo gulo*)," http://srd.alberta.ca/FishWildlife /SpeciesAtRisk/SpeciesSummaries/documents/Wolverine_May_03.pdf, accessed May 5, 2013.

30. IUCN, "IUCN Red List Wolverine," www.iucnredlist.org/details/9561/0, accessed April 26, 2013.

31. Copeland et al., "The Bioclimatic Envelope of the Wolverine," 233–46.

32. Intergovernmental Panel on Climate Change, *Fourth Assessment Report: Climate Change 2007* (Cambridge, UK: Cambridge University Press, 2007).

33. Copeland et al., "The Bioclimatic Envelope of the Wolverine"; Jedediah F. Brodie and Eric Post, "Nonlinear Responses of Wolverine Populations to Declining Winter Snowpack," *Population Ecology* 52 (2010): 279–87.

34. Chris Cegleski et al., "Genetic Diversity and Population Structure of Wolverine (*Gulo gulo*) Populations at the Southern Edge of their Current Distribution in North America with Implications for Genetic Viability," *Conservation Genetics* 7 (2006): 197–211.

35. Chadwick, *The Wolverine Way*, 137–77.

36. Jodi A. Hilty, William Z. Lidicker Jr., and Adina M. Merenlender, *Corridor Ecology: The Science and Practice of Linking Landscapes for Biodiversity Conservation* (Washington, DC: Island Press, 2006), 54–60.

37. Ibid., 100–111.

38. Robert M. Inman et al., "Wolverine Reproductive Rates and Maternal Habitat in Greater Yellowstone," in *Greater Yellowstone Wolverine Program, Cumulative Report May 2007* (Ennis, MT: Wildlife Conservation Society, 2007), 65–84; Mark L. Packila et al., "Wolverine Food Habits in Greater Yellowstone," in *Greater Yellowstone Wolverine Program, Cumulative Report May 2007* (Ennis, MT: Wildlife Conservation Society, 2007), 121–28; Mark L. Packila et al., "Wolverine Road Crossings in Western Greater Yellowstone," in *Greater Yellowstone Wolverine Program, Cumulative Report May 2007* (Ennis, MT: Wildlife Conservation Society, 2007), 103–20.

39. Robert M. Inman, *Wolverine Ecology and Conservation in the Western United States* (dissertation, Swedish University of Agricultural Sciences, Uppsala, 2013), 145 pp.

40. Copeland and Yates, *Wolverine Population Assessment in Glacier National Park.*

41. Anthony Clevenger, interview by Cristina Eisenberg, October 9, 2012, Canmore, AB.

42. Chadwick, *The Wolverine Way*, 269.

Chapter 7: Lynx

1. Mark O'Donoghue et al., "Behavioral Responses of Coyotes and Lynx to the Snowshoe Hare Cycle," *Oikos* 82, no. 1 (1998): 169–83 (describes lynx ambush strategies for hunting snowshoe hares).

2. Douglas A. Maguire, Charles B. Halpern, and David L. Phillips, "Changes in Forest Structure Following Variable-Retention Harvests in Douglas Fir–Dominated Forests," *Forest Ecology and Management* 242, no. 2 (2007): 708–26.

3. John R. Squires et al., "Seasonal Resource Selection of Canada Lynx in Managed Forests of the Northern Rocky Mountains," *Journal of Wildlife Management* 74, no. 8 (2010): 1648–60.

4. USFWS, "Determination of Threatened Status for the Contiguous US Distinct Population Segment of the Canada Lynx and Related Rule; Final Rule," CFR Part 17, *Federal Register* 65, no. 58 (March 24, 2000): 16052; USDA Forest Service, *Northern Rockies Lynx Amendment Draft Environmental Impact Statement* (Missoula, MT: USDA Forest Service, 2004); Mathieu Basille et al., "Selecting Habitat to Survive: The Impact of Road Density on Survival in a Large Carnivore," *PLoS ONE* 8, no. 7 (2013): e65493. doi:10.1371/journal.pone.0065493.

5. Hal Salwasser, "Ecosystem Management: A New Perspective for National Forests and Grasslands," in *Ecosystem Management: Adaptive Strategies for Natural Resources Organizations in the Twenty-First Century* (Levittown, PA: Taylor and Francis, 1999), 85–96.

6. Leonard F. Ruggiero et al., "Species Conservation and Natural Variation among Populations," in *Ecology and Conservation of Lynx in the United States,* ed. Leonard F. Rugiero et al. (Boulder, CO: University Press of Colorado, 2000), 109.

7. Adolph Murie, *A Naturalist in Alaska* (Tucson, AZ: University of Arizona Press, 1961), 14–17.

8. Garth Mowat, Kim G. Poole, and Mark O'Donoghue, "Ecology of Lynx in Northern Canada and Alaska," in *Ecology and Conservation of Lynx in the United States,* 267.

9. John R. Squires et al., "Hierarchical Den Selection of Canada Lynx in Western Montana," *Journal of Wildlife Management* 72, no. 7 (2008): 1497–1506.

10. John R. Squires and Leonard F. Ruggiero, "Winter Prey Selection of Canada Lynx in Northwestern Montana," *Journal of Wildlife Management* 71, no. 2 (2007): 310–15.

11. Paul C. Griffin and L. Scott Mills, "Sinks without Borders: Snowshoe Hare Dynamics in a Complex Landscape," *Oikos* 118 (2009): 1487–98.

12. Benjamin T. Maletzke et al., "Habitat Conditions Associated with Lynx Hunting Behavior during Winter in Northern Washington," *Journal of Wildlife Management* 72, no. 7 (2008): 1473–78.

13. Charles Elton, *Animal Ecology,* 3d edition (Chicago: University of Chicago Press, 2002), 135; Murie, *A Naturalist in Alaska,* 14–28; Stan Boutin et al., "Population Changes of the Vertebrate Community during a Snowshoe Hare Cycle in Canada's Boreal Forest," *Oikos* 74 (1995): 69–80; Charles J. Krebs et al., "Impact of Food and Predation on the Snowshoe Hare Cycle," *Science* 269, no. 5227 (1995): 1112–15.

14. Garth Mowat, Brian G. Slough, and Stan Boutin, "Lynx Recruitment during a Snowshoe Hare Population Peak and Decline in Southwest Yukon," *Journal of Wildlife Management* 60, no. 2 (1996): 441–52.

15. Karen E. Hodges, "Ecology of Snowshoe Hares in Southern Boreal and Montane Forests," in *Ecology and Conservation of Lynx in the United States,* 163–206;

Dennis L. Murray, Todd D. Steury, and James D. Roth, "Assessment of Canada Lynx Research and Conservation Needs in Southern Range: Another Kick at the Cat," *Journal of Wildlife Management* 72, no. 7 (2008): 1463–72.

16. Kevin S. McKelvey, Keith B. Aubry, and Yvette K. Ortega, "History and Distribution of Lynx in the Contiguous United States," in *Ecology and Conservation of Lynx in the United States,* 207–64.

17. Steven W. Buskirk, Leonard F. Ruggiero, and Charles J. Krebs, "Habitat Fragmentation and Interspecific Competition," in *Ecology and Conservation of Lynx in the United States,* 83–100; Gary M. Koehler et al., "Habitat Fragmentation and the Persistence of Lynx Populations in Washington State," *Journal of Wildlife Management* 72, no. 7 (2008): 1518–24; John L. Weaver, *The Transboundary Flathead: A Critical Landscape for Carnivores in the Rocky Mountains,* Working Paper No. 18 (Bozeman, MT: Wildlife Conservation Society, 2001).

18. Rugiero et al., *Ecology and Conservation of Lynx in the United States.*

19. USFWS, "Revised Designation of Critical Habitat for the Contiguous United States Distinct Population Segment of the Canada Lynx; Final Rule," CFR Part 17, *Federal Register* 74, no. 36 (February 25, 2009): 8615.

20. Tanya M. Shenk, "Post-Release Monitoring of Lynx Reintroduced to Colorado" (Fort Collins, CO: Colorado Parks and Wildlife, 2009), 1–55.

21. Jacob S. Ivan, "Monitoring Canada Lynx in Colorado Using Occupancy Estimation: Initial Implementation in the Core Lynx Area" (Fort Collins, CO: Colorado Parks and Wildlife, 2011), 10–20; Michael K. Schwartz et al., "DNA Reveals High Dispersal Synchronizing the Population Dynamics of Canada Lynx," *Nature* 415, no. 31 (2002): 520–22.

22. University of Montana, "Climate Change Hurting Hares: White Snowshoe Hares Can't Hide on Brown Earth," *Science Daily,* March 6, 2009, www.sciencedaily.com /releases/2009/02/090224220347.html, accessed July 6, 2013.

23. Laura Prugh, interview by Cristina Eisenberg, July 16, 2013, Fairbanks, AK

24. Gary R. Bortolotti, "Natural Selection and Coloration: Protection, Concealment, Advertisement, or Deception?" in Geoffrey E. Hill and Kevin J. McGraw, eds., *Bird Coloration,* vol. 2 (Cambridge, MA: Harvard University Press, 2006), 3–35; L. Scott Mills, Marketa Zimova, Jared Oyler, Steven Running, John T. Abatzoglou, and Paul M. Lukacs, "Camouflage Mismatch in Seasonal Coat Color Due to Decreased Snow Duration," *Proceedings of the National Academy of Sciences* 110, no. 18 (2013): 7360–65.

25. Laura Prugh, interview by Cristina Eisenberg, July 16, 2013, Fairbanks, AK.

26. McKenzie et al., "Climatic Change, Wildfire, and Conservation;" John R. Squires et al., "Combining Resource Selection and Movement Behavior to Predict Corridors for Canada Lynx at their Southern Range Periphery," *Biological Conservation* 157 (2013): 187–95.

27. Laura Prugh et al., "The Rise of the Mesopredator," *BioScience* 59 (2009): 779–91; William J. Ripple et al., "Can Restoring Wolves Aid in Lynx Recovery?" *Wildlife Society Bulletin* 35, no. 4 (2011): 514–18.

28. John W. Laundré, Lucina Hernandez, and Kelly B. Altendorf, "Wolves, Elk, and Bison: Re-Establishing the 'Landscape of Fear' in Yellowstone National Park," *Canadian Journal of Zoology* 79 (2001): 1401–9.

29. Ripple et al., "Can Restoring Wolves Aid in Lynx Recovery?"

30. John R. Squires et al., "Missing Lynx and Trophic Cascades in Food Webs: A Reply to Ripple et al.," *Wildlife Society Bulletin* 36, no. 3 (2012): 567–71; Jarod Merkle, Daniel R. Stahler, and Douglas W. Smith, "Interference Competition between Gray Wolves and Coyotes in Yellowstone National Park," *Canadian Journal of Zoology* 87 (2009): 56–63; Hodges, "Ecology of Snowshoe hares in Southern Boreal and Montane Forests," 163–206.

31. Squires et al., "Missing Lynx and Trophic Cascades in Food Webs."

32. Laura Prugh, interview by Cristina Eisenberg, July 16, 2013, Fairbanks, AK.

33. Ripple et al., "Can Restoring Wolves Aid in Lynx Recovery?"

34. USFWS, "Revised Designation of Critical Habitat for the Contiguous United States Distinct Population Segment of the Canada Lynx; Final Rule," CFR Part 17, *Federal Register* 74, no. 36 (February 25, 2009): 8615; USDA Forest Service, *Northern Rockies Lynx Amendment* (2004).

Chapter 8: Cougar

1. William J. Ripple and Robert L. Beschta, "Trophic Cascades in Yellowstone: The First 15 Years after Wolf Reintroduction," *Biological Conservation* 145 (2011): 205–13.

2. Aldo Leopold, *A Sand County Almanac: And Sketches Here and There*, 2d ed. (Oxford: Oxford University Press, 1968).

3. Kristin Nowell and Peter Jackson, *Wild Cats: Status Survey and Conservation Action Plan* (Cambridge, UK: IUCN, 1996), 131.

4. Melanie Culver, "Lessons and Insights from Evolution, Taxonomy, and Conservation Genetics," in *Cougar Ecology and Conservation*, ed. Maurice Hornocker and Sharon Negri (Chicago: University of Chicago Press, 2010), 28–30.

5. Steve Pavlick, "Sacred Cat," in *Listening to Cougar*, ed. Mark Bekoff and Cara Blessley Low (Boulder, CO: University of Colorado Press, 2007).

6. Jaguars can be larger, but within the Carnivore Way jaguars and cougars can be similar in size; see: Robert H. Bush, *The Cougar Almanac* (Guilford, CT: Lyons Press, 2004), 26–33.

7. Ibid.

8. Kenneth A. Logan and Linda L. Sweanor, "Behavior and Social Organization of a Solitary Carnivore," in *Cougar Ecology and Conservation*, 105-17.

9. John W. Laundré and Lucina Hernandez, "The Amount of Time Female Pumas *Puma concolor* Spend with Their Kittens," *Wildlife Biology* 14, no. 2 (2008): 221–27.

10. Becky M. Pierce et al., "Migratory Patterns of Mountain Lions: Implications for Social Regulation and Conservation," *Journal of Mammalogy* 80, no 3 (1999): 986–92.

11. Linda L. Sweanor, Kenneth A. Logan, and Maurice G. Hornocker, "Cougar Dispersal Patterns, Metapopulation Dynamics, and Conservation," *Conservation Biology* 14, no. 3 (2000): 798–809.

12. Cougar Management Guidelines Working Group, *Cougar Management Guidelines* (Bainbridge Island, WA: WildFutures, 2005), 11–25; Nowell and Jackson, *Wild Cats*, 131.

13. Ibid.

14. John W. Laundré, "Summer Predation Rates on Ungulate Prey by a Large Keystone Predator: How Many Ungulates Does a Large Predator Kill?" *Journal of Zoology* 275 (2008): 341–48; Kenneth A. Logan and Linda L. Sweanor, *Desert Puma: Evolutionary Ecology and Conservation of an Enduring Carnivore* (Washington, DC: Island Press, 2001), 322–27; John W. Laundré, Lucina Hernandez, and Susan G. Clark, "Impact of Puma Predation on the Decline and Recovery of a Mule Deer Population in Southeastern Idaho," *Canadian Journal of Zoology* 84 (2006): 1555–65.

15. Paul Beier, "A Focal Species for Conservation," in *Cougar Ecology and Conservation,* 178–79; Logan and Sweanor, *Desert Puma,* 341–58.

16. Cougar Management Guidelines Working Group, *Cougar Management Guidelines,* 63–70.

17. Tony K. Ruth and Kerry Murphy, "Cougar–Prey Relationships," in *Cougar Ecology and Conservation,* 138–62; Kyran E. Kunkel et al., "Winter Prey Selection by Wolves and Cougars in and near Glacier National Park, Montana," *Journal of Wildlife Management* 63 (1999): 901–10; Mark Elbroch, *Annual Report, Garfield-Mesa Lion Project* (2012).

18. Cougar Management Guidelines Working Group, *Cougar Management Guidelines,* 63–70.

19. John W. Laundré, *Phantoms of the Prairie: The Return of Cougars to the Midwest* (Madison, WI: The University of Wisconsin Press, 2012); John W. Laundré and Lucina Hernandez, "Winter Hunting Habitat of Pumas *Puma concolor* in Northwestern Utah and Southern Idaho," *Wildlife Biology* 9, no. 2 (2003):123–29.

20. Paul Beier et al., "Mountain Lion (*Puma concolor*)," in *Urban Carnivore Ecology,* ed. Stanley D. Gehrt, Seth P. D. Riley, and Brian L. Cyper (Baltimore, MD: Johns Hopkins University Press, 2010), 177–89; Brian N. Kertson et al., "Cougar Space Use and Movements in the Wildland-Urban Landscape of Western Washington," *Ecological Applications* 21, no. 8 (2011): 2866–81; John W. Laundré et al., "Evaluating Potential Factors Affecting Puma *Puma concolor* abundance in the Mexican Chihuahuan Desert," *Wildlife Biology* 15 (2009): 207–12.

21. Kristin Nowell and Peter Jackson, *Wild Cats*, 131.

22. Laundré, *Phantoms of the Prairie*; Daniel Thompson and John A. Jenks, "Dispersal Movements of Subadult Cougars from the Black Hills: The Notions of Range Expansion and Recolonization," *Ecosphere* 1, no. 4 (2010): 1–11; Michelle A. Larue, "Cougars Are Recolonizing the Midwest: Analysis of Cougar Confirmations during 1990–2008," *Journal of Wildlife Management* 76, no. 7 (2012): 1364–69.

23. Sue Morse, "Keeping Track," http://keepingtrack.org/, accessed June 30, 2013; UCSC Puma Project, http://santacruzpumas.org/, accessed June 30, 2013.

24. Harley G. Shaw et al., *Puma Field Guide* (New Mexico: Puma Network, 2007), 17.

25. Cougar Management Guidelines Working Group, *Cougar Management Guidelines*, 5–9.

26. Arturo Caso et al., *Puma concolor*, in *IUCN Red List of Threatened Species*, Version 2012.2, www.iucnredlist.org, accessed June 15, 2013.

27. Catherine M. S. Lambert et al., "Cougar Population Dynamics and Viability in the Pacific Northwest," *Journal of Wildlife Management* 70, no. 1 (2006): 2456–64; Hilary S. Cooley et al., "Source Populations in Carnivore Management: Cougar Demography and Emigration in a Lightly Hunted Population," *Animal Conservation* 12 (2009): 321–28.

28. Philip A. Stephens, William J. Sutherland, and Robert P. Freckleton, "What Is the Allee Effect?" *Oikos* (1999): 185–90; Paul Beier, "Dispersal of Juvenile Cougars in Fragmented Habitat," *Journal of Wildlife Management* 59, no. 2 (1995): 228–37; Paul Beier, "Determining Minimum Habitat Areas and Habitat Corridors for Cougars," *Conservation Biology* 7, no. 1 (1993): 94–108.

29. Hugh S. Robinson et al., "Sink Populations in Carnivore Management: Cougar Demography and Immigration in a Hunted Population," *Ecological Applications* 18, no. 4 (2008): 1028–37; John W. Laundré and Susan G. Clark, "Managing Puma Hunting in the Western United States: Through a Metapopulation Approach," *Animal Conservation* 6 (2003): 159–70.

30. Kevin R. Crooks and Michael E. Soulé, "Mesopredator Release and Avifaunal Extinctions in a Fragmented System," *Nature* 400 (1999): 563–66.

31. William J. Ripple and Robert L. Beschta, "Linking Cougar Decline, Trophic Cascades, and Catastrophic Regime Shift in Zion National Park," *Biological Conservation* 133 (2006): 397–408.

32. William J. Ripple and Robert L. Beschta, "Trophic Cascades Involving Cougar, Mule Deer, and Black Oaks in Yosemite National Park," *Biological Conservation* 141 (2007): 1249–56.

33. Rolf Peterson, interview by Cristina Eisenberg, April 29, 2008, Michigan Technological University, Houghton, MI; Doug Smith, interview by Cristina Eisenberg, November 4, 2008, Yellowstone National Park.

34. Cristina Eisenberg, *Complexity of Food Web Interactions in a Large Mammal System* (dissertation, Oregon State University, Corvallis, OR, 2012).

35. L. Mark Elbroch and Heiko Wittmer, "Table Scraps: Inter-Trophic Food Provisioning by Pumas," *Biology Letters* 8, no. 5 (2012): 776–79.

36. Beier, "A Focal Species for Conservation," 177–78; Kyran Kunkel et al., "Assessing Wolves and Cougars as Conservation Surrogates," Paper 1157, USDA National Wildlife Research Center, Staff Publications (2013).

37. Paul Beier, "Cougar Attacks on Humans in the United States and Canada," *Wildlife Society Bulletin* 19, no. 4 (1991): 403–12; Linda L. Sweanor and Kenneth A. Logan, "Cougar–Human Interactions," in *Cougar Ecology and Conservation*, 190–205.

38. David Mattson, Kenneth Logan, and Linda Sweanor, "Factors Governing Risk of Cougar Attacks on Humans," *Human–Wildlife Interactions* 5, no. 1 (2011): 135–58.

39. Steven Michel, interview by Cristina Eisenberg, October 9, 2012, Banff National Park, AB.

40. Joel Berger, "Fear, Human Shields, and the Redistribution of Prey and Predators in Protected Areas," *Biology Letters* 3, no. 6 (2007): 620–23.

41. David J. Mattson and Susan G. Clark, "The Discourses of Incidents: Cougars on Mt. Elden and in Sabino Canyon, Arizona," *Policy Sci* 45 (2012): 315–43; David J. Mattson and Susan G. Clark, "People, Politics, and Cougar Management," in *Cougar Ecology and Conservation*, 190–205.

42. Beier, "Dispersal of Juvenile Cougars in Fragmented habitat;" Beier, "A Focal Species for Conservation," 188–89; Claire C. Gloyne and Anthony P. Clevenger, "Cougar *Puma concolor* Use of Wildlife Crossing Structures on the Trans-Canada Highway in Banff National Park, Alberta," *Wildlife Biology* 7, no. 2 (2001): 117–24; Christopher C. Wilmers et al., "Scale-Dependent Behavioral Responses to Human Development by a Large Predator, the Puma," *Plos One* 8, no. 4 (2013): 1–11.

43. Tucker Murphy and David W. Macdonald, "Pumas and People: Lessons in the Landscape of Tolerance from a Widely Distributed Field," in *Biology and Conservation of Wild Felids*, ed. David W. Macdonald and Andrew J. Loveridge (Oxford: Oxford University Press, 2010), 431–52.

Chapter 9: Jaguar

1. David E. Brown and Carlos A. Lopez Gonzalez, *Borderland Jaguars* (Salt Lake City, UT: University of Utah Press, 2001), 1–3.

2. Ibid., 125–28.

3. Kristin Nowell and Peter Jackson, *Wild Cats: Status Survey and Conservation Action Plan* (Cambridge, UK: IUCN, 1996), 118–19.

4. Brown and Lopez Gonzalez, *Borderland Jaguars*, 29–32.

5. Ibid., 4–5.

6. Nowell and Jackson, *Wild Cats,* 118–19; Emil B. McCain and Jack L. Childs, "Evidence of Resident Jaguars (*Panthera onca*) in the Southwestern United States and the Implications for Conservation," *Journal of Mammalogy* 89, no. 1 (2008): 1–10.

7. Alan Rabinowitz, *Jaguar: One Man's Struggle to Establish the World's First Jaguar Preserve* (Washington, DC: Island Press, 2000), 8; Alan Rabinowitz, "The Present Status of Jaguars (*Panthera onca*) in the Southwestern United States," *Southwestern Naturalist* 44, no. 1 (1999): 96–100.

8. David E. Brown, ed., *Biotic Communities: Southwestern United States and Northwestern Mexico* (Salt Lake City, UT: University of Utah Press, 1994), 101–4.

9. Since most jaguars occur in Central and South America, where they tend to be larger, this makes the average size of the species significantly larger than cougar average size, despite the fact that the northern jaguar is similar in size to the cougar.

10. Brown and Lopez Gonzalez, *Borderland Jaguars,* 21–25.

11. Kevin L. Seymour, "*Panthera onca,*" *Mammalian Species* 340 (1989): 1–9.

12. Nowell and Jackson, *Wild Cats,* 118–19.

13. Harley Shaw, "The Emerging Cougar Chronicle," in *Cougar Ecology and Conservation,* 17–26.

14. Carlos A. Lopez Gonzalez and Brian J. Miller, "Do Jaguars (*Panthera onca*) Depend on Large Prey?" *Western North American Naturalist* 62, no. 2 (2002): 218–22; Fernando Cesar Cascelli de Azevedo, "Food Habits and Livestock Depredation of Sympatric Jaguars and Pumas in the Iguaçu National Park Area, South Brazil," *Biotropica* 40, no. 4 (2008): 494–500.

15. Lane Simonian, *Defending the Land of the Jaguar: A History of Conservation in Mexico* (Austin, TX: University of Texas Press, 1995), 154–56; Eric W. Sanderson et al., "Planning to Save a Species: The Jaguar as a Model," *Conservation Biology* 16, no. 1 (2002): 58–72.

16. Rabinowitz, "The Present Status of Jaguars (*Panthera onca*) in the Southwestern United States."

17. Arturo Caso et al., *Panthera onca,* in *IUCN Red List of Threatened Species,* Version 2013.1, www.iucnredlist.org, accessed on August 3, 2013.

18. Brown and Lopez Gonzalez, *Borderland Jaguars,* 138–40; USFWS, "Designation of Critical Habitat for Jaguar."

19. Rabinowitz, "The Present Status of Jaguars (*Panthera onca*) in the Southwestern United States"; Kurt A. Menke and Charles L. Hayes, "Evaluation of the Relative Suitability of Potential Jaguar Habitat in New Mexico" (Santa Fe, NM: New Mexico Department of Game and Fish, 2003); James R. Hatten et al., "A Spatial Model of Potential Jaguar Habitat in Arizona," *Journal of Wildlife Management* 69, no. 3 (2005): 1024–33.

20. Melissa Grigione et al., "Neotropical Cats in Southeast Arizona and Surrounding Areas: Past and Present Status of Jaguars, Ocelots, and Jaguarundis," *Mastozoologia Neotropical* 14, no. 2 (2007): 189–99; Octavio C. Rosas-Rosas and Louis C. Bender, "Population Status of Jaguars (*Panthera onca*) and pumas (*Puma concolor*) in Northeastern Sonora, Mexico," *Acta Zoologica Mexicana* 28, no. 1 (2012): 86–101.

21. Alan Rabinowitz and Kathy A. Zeller, "A Range-Wide Model of Landscape Connectivity and Conservation for the Jaguar," *Biological Conservation* 143 (2010): 939–45, Alan Rabinowitz, *An Indomitable Beast: The Remarkable Journey of the Jaguar* (Washington, DC: Island Press, 2014, *in press*).

22. Randall C. Archibold, "Border Plan Will Address Harm Done at Fence Site," *New York Times*, January 16, 2009, www.nytimes.com/2009/01/17/us/17border .html?ref=borderfenceusmexico, accessed December 15, 2012; Melissa Grigione et al., "Identifying Potential Conservation Areas for Felids in the USA and Mexico: Integrating Reliable Knowledge across an International Border," *Oryx* 43, no. 1 (2009): 78–86; Octavio Rosas-Rosas et al., "Habitat Correlates of Jaguar Kill-Sites of Cattle in Northeastern Sonora, Mexico," *Human-Wildlife Interactions* 4, no. 1 (2010): 103–11.

23. Aldo Leopold, *A Sand County Almanac: And Sketches Here and There*, 2d ed. (Oxford: Oxford University Press, 1968), 143.

24. Brian Miller et al., "Using Focal Species in the Design of Nature Reserve Networks," *Wild Earth* Winter (1998/99): 81–92.

25. Sandra M. C. Cavalcanti and Eric M. Gese, "Kill Rates and Predation Patterns of Jaguars (*Panthera onca*) in the Southern Pantanal, Brazil," *Journal of Mammalogy* 91, no. 3 (2010): 722–36.

26. Brown and Lopez Gonzalez, *Borderland Jaguars*, 67–77; Rabinowitz, "The Present Status of Jaguars (*Panthera onca*) in the Southwestern United States."

27. Will Rizzo, "Return of the Jaguar," *Smithsonian*, December 2005; Jeremy Voas, "Cat Fight on the Border," *High Country News*, October 15, 2007; Janay Brun, interview by Cristina Eisenberg, August 11, 2013.

28. Janay Brun, interview by Cristina Eisenberg, August 11, 2013.

29. Ibid.

30. Dennis Wagner, "Cover-Up amid Celebrations," *Arizona Republic*, December 10, 2012.

31. AZGFD information on the Macho B investigation can be found on the Arizona Game and Fish Department website, http://www.azgfd.gov/w_c/jaguar/MachoB .shtml, accessed August 24, 2013.

32. Tony Davis, "Was Arizona Jaguar Macho B's Euthanasia Unnecessary?" *Arizona Star*, March 31, 2009.

33. Janay Brun, interview by Cristina Eisenberg, August 11, 2013.

34. Dennis Wagner, "Cover-Up amid Celebrations," *Arizona Republic*, December 10, 2012.

35. Dennis Wagner, "Web of Intrigue Surrounds Death of Jaguar Macho B," *Arizona Republic*, December 12, 2012.

36. Janay Brun, "Truth or Consequence," *Three Coyotes* 1, no. 2 (2011).

37. Hatten et al., "A Spatial Model of Potential Jaguar Habitat in Arizona."

38. Menke and Hayes, "Evaluation of the Relative Suitability of Potential Jaguar Habitat in New Mexico."

39. Michael Robinson, "Suitable Habitat for Jaguars in New Mexico" (Tucson, AZ: Center for Biological Diversity, 2005).

40. Jim Nintzel, "Jaguar Spotted in the Santa Rita Mountains," *Tucson Weekly*, December 20, 2012.

41. USFWS, "Preliminary Strategy for Jaguar Recovery Is Completed" (Tucson, AZ: USFWS, April 19, 2012).

42. USFWS, "Designation of Critical Habitat for Jaguar"; Grigione et al., "Neotropical Cats in Southeast Arizona and Surrounding Areas."

43. Alan Rabinowitz, "Jaguars Don't Live Here Anymore," *New York Times*, January 25, 2010; Rabinowitz and Zeller, "A Range-Wide Model of Landscape Connectivity and Conservation for the Jaguar."

44. Michael J. Robinson, "Comment to USFWS on Proposed Rule to Designate Critical Habitat for the Jaguar," Center for Biological Diversity, October 12, 2012.

45. Kurt Menke, interview by Cristina Eisenberg, August 7, 2013.

46. Rabinowitz, *Jaguar: One Man's Struggle to Establish the World's First Jaguar Preserve*, xiv.

47. Grigione et al., "Identifying Potential Conservation Areas for Felids in the USA and Mexico"; Northern Jaguar Project, www.northernjaguarproject.org/, accessed August 3, 2013.

48. Kathy Zeller, *Jaguars in the New Millennium Data Set Update: The State of the Jaguar in 2006* (New York: Wildlife Conservation Society, 2007).

49. Octavio C. Rosas-Rosas and Raul Valdez, "The Role of Landowners in Jaguar Conservation in Sonora, Mexico," *Conservation Biology* 24, no. 2 (2010): 366–71.

50. Tony Davis, "Mexican Rancher Struggles to Shift from Cattle to Conservation," *High Country News*, April 20, 2012; Cynthia Lee Wolf, interview by Cristina Eisenberg, Columbia Falls, MT, August 8, 2013.

Conclusion: Earth Household

1. Edward O. Wilson, *The Diversity of Life* (Cambridge, MA: Belknap Press of Harvard University Press, 1992).

2. Michael E. Soulé and John Terborgh, *Continental Conservation: Scientific Foundations of Regional Reserve Networks* (Washington, DC: Island Press, 1999).

3. Michael P. Nelson and John A. Vucetich, "Environmental Ethics and Wildlife Management," *Human Dimensions of Wildlife Management*, ed. Daniel J. Decker, Shawn J. Riley, and William F. Siemer (New York: Johns Hopkins University Press, 2012), 223–37.

4. Ibid.; Aldo Leopold, *A Sand County Almanac* (New York: Ballantine, 1986), 240, 262.

5. Peter Matthiessen, *Wildlife in America*, 3d ed. (New York: Viking, Compass, 1967); Thomas R. Dunlap, *Saving America's Wildlife* (Princeton, NJ: Princeton University Press, 1988).

6. Thomas R. Dunlap, *Saving America's Wildlife* (Princeton, NJ: Princeton University Press, 1988), 10–14, 38.

7. Michael P. Nelson et al., "An Inadequate Construct? North American Model: What's Flawed, What's Missing, What's Needed," *Wildlife Professional* (Summer 2011): 58–60.

8. Valerius Geist, Shane P. Mahoney, and John F. Organ, "Why Hunting Has Defined the North American Model of Wildlife Conservation," *Transactions of the North American Wildlife and Natural Resources Conference* 66 (2001): 175–83.

9. Charles List, *Hunting, Fishing, and Environmental Virtue* (Corvallis, OR: Oregon State University Press, 2013); Steven J. Holmes, *The Young John Muir: An Environmental Biography* (Madison, WI: University of Wisconsin Press, 1999).

10. Michael P. Nelson, interview by Cristina Eisenberg, Corvallis, Oregon, April 23, 2013; John A. Vucetich, interview by Cristina Eisenberg, Corvallis Oregon, April 16, 2013; Nelson et al., "An Inadequate Construct."

11. Estes et al., "Trophic Downgrading of Planet Earth," *Science* 33, no. 6040 (2011): 301–6.

12. J. Baird Callicott, *Companion to* A Sand County Almanac: *Interpretive and Critical Essays* (Madison, WI: University of Wisconsin Press, 1987).

13. Dave Foreman, *Rewilding North America: A Vision for Conservation in the 21st Century* (Washington, DC: Island Press, 2004).

14. List, *Hunting, Fishing, and Environmental Virtue.*

15. Aldo Leopold, *Round River*, ed. Luna B. Leopold (Oxford: Oxford University Press, 1993).

16. Robert Jonathan Cabin, *Intelligent Tinkering: Bridging the Gap between Science and Practice* (Washington, DC: Island Press, 2011).

17. Gary Snyder, *Earth House Hold* (New York: New Directions, 1969).

Glossary

adaptive management   A science-based approach to resource management that acknowledges uncertainty and that views policy and management decisions as testable hypotheses, to be revised in light of new information.

additive mortality   A situation where harvest by humans (e.g., through hunting of a species) will cause a population of that species to decline in greater numbers than it would if it were not harvested.

Anthropocene   The geological epoch that began with the Industrial Revolution in the late seventeenth century; characterized by human modification of ecosystems, a growing human population, and a growing extinction rate. Also called the Age of Man.

background extinction   The normal rate of extinction, usually one to five species per year.

biodiversity   The variety of living organisms at all levels of organization, including genetic, species, and higher taxonomic levels; the variety of habitats and ecosystems, as well as the processes occurring therein.

biome   A complex biotic community characterized by distinctive plant and animal species and maintained under the climatic condition of the region.

biomass   The total mass of living matter within a given area.

biota   Plants and animals, comprising all the organisms in an ecosystem.

biotic   Relating to or caused by living organisms.

boreal   Relating to the northern regions; pertaining to the northern forests, below the Arctic Circle.

bottom-up control   Regulation of food web components in an ecosystem by either primary producers (plants) or by limits to the input of nutrients.

browsing   Feeding on the leaves, branches, or shoots of woody plant species such as shrubs and trees.

carrying capacity   In reference to animal populations, the maximum population level that can be supported by habitat without damage to the population or deterioration of the fitness of the species (e.g., through malnutrition, increase in disease). Population ecologists also refer to this as *K*. There are also *social* and *economic* carrying capacities, which are the population densities a species can achieve in places subject to multiple human land uses.

circumboreal   Throughout all boreal regions worldwide.

climate   Weather conditions over time, 30 years being a typical increment for measurements.

climate change   Change in the mean of one or more measures of climate (e.g., precipitation, temperature) over an extended period of time, typically decades.

community   In ecology, this refers to a group of different types of organisms that coexist. Additionally, a community is usually characterized by its dominant plant species (e.g., an "aspen community"). Community types are formally based on dominant tree species, or, in non-forested areas, on dominant grass species. Ecologists defined and standardized these community types in the 1920s and 1930s.

community ecology   The study of the structure of communities and how they vary in time and space in response to physical and biotic factors.

compensatory mortality   A situation where harvest by humans (e.g., through hunting of a species) does not cause the population of that species to decline further than it normally would if it were not harvested.

competition   A non-trophic (food web) interaction between two species that has negative consequences for both.

connectivity   A measure of the ability of organisms to move among separated patches of suitable habitat; can be viewed at various spatial scales.

consumers   A term used in community ecology for herbivores.

control   A group of experimental subjects or an area not exposed to the treatment being applied or investigated, to be used for comparison.

corridor An area of habitat that connects wildlife populations separated by human activities such as roads, natural-resources extraction, or development, and thus allows for genetic exchange between these separate populations.

critical habitat According to the Endangered Species Act, the ecosystems upon which endangered and threatened species depend for survival.

delisting The process by which US Fish and Wildlife Service removes a species from the list of threatened and endangered plant and animal species.

deme A local population, part of a metapopulation.

density-dependent factors Factors whose effects on a population change in relative intensity as population density changes, such as factors that affect the birth rate or mortality of a species. Such factors typically include predation.

distinct population segment According to the Endangered Species Act, a population of a species designated for protection.

dominant species Species whose influences result from their great abundance and account for most of the biomass in a community, and are thus primary components of community structure.

ecological integrity The ability of an ecosystem to self-correct after a disturbance and return to the state normal for that system.

ecological restoration Assisting the recovery of an ecosystem that has been degraded, damaged, or destroyed. The goal of this process is to emulate the structure, function, diversity, and dynamics of the specified ecosystem, but not necessarily to return it to an earlier condition.

ecological extinction The reduction of a species to such low abundance that, although still present in the community, it no longer interacts significantly with other species.

ecologically effective population In conservation biology, a population of a *keystone species* of sufficient density and distribution to cause a trophic cascade; in population ecology, the breeding females of a species in a population.

ecology The study of biotic communities, the living organisms they contain, and these organisms' abundance and relationships with each other and with their environment.

ecoregion A relatively large area of land or water that contains a geographically distinct assemblage of natural communities.

ecotone The border between two different ecosystems or habitat types, such as where field and forest meet.

empirical Derived from experiment and observation rather than theory; based on data.

endemic   Any localized process or pattern, but usually applied to a highly localized or restrictive geographic distribution of a species.

fecundity   Potential reproductive ability of a species or a population; functions as a measure of the fitness of a species or population.

feedback   A phenomenon in which a system's output modifies input to the system, thus becoming self-perpetuating in its dynamics. Feedback loops can be positive or negative. In a positive feedback loop, for example, the presence of a specific pollinator that only pollinates a particular plant species increases the amount of that plant species, which then increases the population of that pollinator.

flagship species   A charismatic species with broad popular appeal. Flagship species can function as ambassadors that advance conservation, because they engage people in caring about conservation. The sea otter (*Enhydra lutris*) is an example of a flagship species.

focal species   A species whose requirements for survival represent factors important to ecosystem function. Scientists and managers pay attention to focal species because, due to budget limitations, they can't pay attention to all species, and also because focal species can represent broad ecological issues.

food web   The structure of observable trophic (food-related) linkages in a community.

forb   A broad-leaved herb that is not a grass, especially one growing in meadow, prairie, or field environments.

fragmentation   The transformation of a continuous habitat into habitat patches that usually vary in size and configuration.

grazing   Feeding on herbaceous plants in a field or pasture.

Green World Hypothesis   A hypothesis suggesting that the world's abundance of plant biomass results from top predators controlling their herbivore prey, via predation, thereby having an indirect effect on vegetation, enabling it to grow.

habitat   The dwelling place of an organism or community, which provides the necessary conditions for its existence.

habitat fragmentation   The disruption of extensive habitats into isolated and small patches; often applied to forested habitats that have been fragmented by agricultural development or logging.

herbaceous plants   Plants whose leaves and stems die down to soil level at the end of the growing season.

herbivore   A plant-eating organism.

herbivory   A form of predation in which an organism, known as an *herbivore*, consumes plants.

heterogeneous Consisting of dissimilar elements; diverse. In ecology, this term is used to refer to landscape, ecosystem, or community structure.

heuristic A teaching method in which students make discoveries on their own, i.e., learning by doing.

indicator species A species that demonstrates how an ecosystem is functioning. Indicator species are usually only present under a certain set of circumstances. Wildlife managers often use indicator species as a shortcut to monitoring a whole ecosystem.

indirect effects The effects of a species on other species, often through trophic interactions; also called *secondary effects.*

intensive management Wildlife management with the objective of reducing the population of a species to the lowest possible level above extinction. Typically applied to large carnivore management to boost prey populations for hunting by humans.

irruption A sudden, explosive increase in population numbers, which will exceed *carrying capacity* if not limited or controlled; often involves an exponential increase in population size in the absence of predation.

keystone species A species whose impacts on its community or ecosystem are large, and much larger than would be expected from the species' abundance.

landscape ecology The branch of ecology that studies the interactions among different biotic community types on a relatively large scale.

listing The process by which US Fish and Wildlife Service puts a species on the list of threatened and endangered plant and animal species.

matrix The various different biotic communities that surround habitat patches, which can influence metapopulations.

megafauna Species with adults weighing more than 100 pounds.

mesopredators Medium-sized predators, such as coyotes, raccoons, and foxes, which often increase in abundance when larger predators are eliminated.

meta-analysis The process of analyzing research in a specific area of inquiry by comparing and combining the results of several earlier independent but related studies.

metapopulation A large population of animals made up of smaller populations (also called *demes*), which ideally can interbreed and move freely among each other's ranges.

Natural Selection A process theorized by Charles Darwin as the principal mechanism for the evolution of species, by which organisms best adapted to their environment survive, and which shapes traits in all organisms.

net primary production Also called NPP, the energy flow in ecosystems and a measure of biomass, or ability of things to grow. Its components are energy flow via sunlight, moisture, and photosynthesis.

niche The position or role of a species in an ecosystem. Popularly defined as a species' "job."

order of magnitude A tenfold change in number that represents one exponential level, plus or minus; a hundredfold change represents a second exponential level, etc.

paradigm An established pattern of thought, often applied to a dominant ecological or evolutionary viewpoint; e.g., during earlier decades, the dominant paradigm in ecology held that biotic communities underwent orderly, predictable development after a catastrophic disturbance such as a volcanic eruption or severe fire.

permafrost Permanently frozen soil, sediment, or rock that remains at or below freezing for at least two years, generally occurring in North America in Alaska, the Yukon, and northern Canada, between the latitudes of 60° N and 68° N. However, permafrost also occurs in patches in the subalpine and alpine zones in mountainous regions at lower latitudes.

permeability With regard to landscape ecology, the ability of a land area to allow the passage of animals.

persistence The capacity of a population to live for 100 years or longer.

phenology The scientific study of flowing, breeding, migration, and other periodic biological phenomena as they relate to climate.

philopatry The propensity of an adult animal to remain near its point of origin, i.e., within its maternal home range.

Pleistocene An epoch of the Quaternary period, also known as the Ice Age, extending from the end of the Pliocene, some 1.64 million years ago, to the beginning of the Holocene, approximately 10,000 years ago.

recruitment Survival of juveniles (plants or animals) for a period sufficient for them to reach adulthood.

recruitment gap Missing age classes in a tree community due to chronic herbivory.

release Habitat expansion or density increase of a species when one or more competing species are absent.

resilience The ability of systems to react to a disturbance by recovering rapidly, while still maintaining the same ecological characteristics and relationships among organisms.

scale   The magnitude of a region or process. Refers to both size, e.g., a relatively small-scale patch or a relatively large-scale landscape, and time, e.g., relatively rapid ecological succession or slow evolution of species.

secondary effects   The effects of population loss of a species upon other species, often through trophic interactions; also called *indirect effects*.

sink   An area where numbers of a particular species may decline sharply, possibly due to a variety of causes, but often due to habitat fragmentation and human development.

source-sink dynamics   A conceptual model used in ecology to describe how changes in habitat quality may affect local population dynamics of species. Good habitat causes the local population of a species to grow, creating a *source*; poor habitat causes the local population of a species to decline, creating a *sink*.

speciation   The process of species formation.

species diversity   The proportional distribution of species, i.e., species abundance. See *species richness*.

species richness   The simplest measure of biodiversity, species richness refers to the number of species in the site being sampled. It does not consider the abundance of individual species.

speciose   Containing many species; high in overall biodiversity.

stochastic   Random; used particularly to describe any random process, such as mortality, or events attributable to extreme weather, disease, or causes beyond human control.

strongly interacting species   Species having a large effect on the other species with which they interact. A communities or ecosystem may have many strong interactors, occurring at all trophic levels. This general term can include *keystone species*.

succession   The natural, sequential change of species composition in a community in a given area.

symbiosis   A close relationships between at least two species.

sympatric   Species that live in the same area at the same time, but do not interbreed.

taiga   Sparse forests characteristic of the northern half of the boreal biome, which are comprised mainly of spruce.

taxa   A group or category of species, such as genus, family, or order.

top-down control   Regulation of lower food web components by an apex predator.

transect   A line through an area being used to sample and monitor organisms or conditions.

trophic   Of or pertaining to food, as in a trophic level of a food web.

trophic cascade   A condition in which the presence of an apex predator causes direct (mortality-driven) and indirect (fear-driven) effects on its primary prey (an herbivore), and that in turn causes this prey species to reduce its impact on the vegetation that it eats, enabling plants to grow and provide habitat for other species.

tundra   The treeless zone between the Arctic Ocean and the beginning of the treeline. The soil is frozen here year-round; consequently, it supports only low-lying vegetation.

umbrella species   A species that requires a large area for its existence; protection of such a species offers protection to other species that share the same habitat.

ungulate   A hoofed mammal.

watershed   The land area draining into a stream; the watershed for a major river may encompass a number of smaller watersheds that ultimately combine.

woody plants   Plants that contain woody fiber (called *lignin*) in their tissues, particularly in their stems. Woody species are usually perennial trees and shrubs.

vagility   The ability of an organism to move; generally applied to species and population dynamics.

About the Author

Cristina Eisenberg conducts trophic cascades research that focuses on wolves in Rocky Mountain ecosystems. She also studies how fire interacts with apex predator effects in forest communities. She holds a PhD in forestry and wildlife from Oregon State University, where she teaches ecological restoration and public policy. She is a Smithsonian Conservation Biology Research Associate, a Boone and Crockett Club Professional Member, and an Aldo Leopold scholar. Dr. Eisenberg has authored multiple peer-reviewed scientific and literary journal articles and several book chapters. Her first book, *The Wolf's Tooth: Keystone Predators, Trophic Cascades, and Biodiversity*, was published in 2010 by Island Press. When she is not teaching, she lives in northwest Montana in a remote valley where the large carnivores outnumber the humans.

Index